SpringerBriefs in Psychology

Psychology and Cultural Developmental Science

Series editors

Giuseppina Marsico, University of Salerno, Salerno, Italy;
Centre for Cultural Psychology, Aalborg University, Aalborg
Denmark

Jaan Valsiner, Centre for Cultural Psychology, Aalborg University,
Aalborg, Denmark

SpringerBriefs in Psychology and Cultural Developmental Science will be an extension and topical completion of *IPBS: Integrative Psychological and Behavioral Science Journal* (Springer, chief editor: Jaan Vasiner) expanding some relevant topics in the form of single (or multiple) authored book. The Series will have a clearly defined international and interdisciplinary focus hosting works on the interconnection between Cultural Psychology and other Developmental Sciences (biology, sociology, anthropology, etc). The Series aims at integrating knowledge from many fields in a synthesis of general science of Cultural Psychology as a new science of the human being.

The Series will include books that offer a perspective on the current state of developmental science, addressing contemporary enactments and reflecting on theoretical and empirical directions and providing, also, constructive insights into future pathways.

Featuring compact volumes of 100 to 115 pages, each Brief in the Series is meant to provide a clear, visible, and multi-sided recognition of the theoretical efforts of scholars around the world.

Both solicited and unsolicited proposals are considered for publication in this series. All proposals will be subject to peer review by external referees.

More information about this series at http://www.springer.com/series/15388

Alan Rayner

The Origin of Life Patterns

In the Natural Inclusion of Space in Flux

 Springer

Alan Rayner
Bath Bio*Art
Bathford, Bath and North East Somerset
UK

ISSN 2192-8363   ISSN 2192-8371   (electronic)
SpringerBriefs in Psychology
ISBN 978-3-319-54605-6   ISBN 978-3-319-54606-3   (eBook)
DOI 10.1007/978-3-319-54606-3

Library of Congress Control Number: 2017932630

Printed on acid-free paper

This Springer imprint is published by Springer Nature
The registered company is Springer International Publishing AG
The registered company address is: Gewerbestrasse 11, 6330 Cham, Switzerland

What May Not Be Obvious
Every body is a cavity at heart
*Every figure reconfigures both in science and
in art*
*Every face is interfacing from no bottom to no
top*
*Every faith is interfaith that cannot tell us
where to stop*
*Every lining opens inwards as it brings its
inside out*
*Every curtain closes outwards to conceal its
inner doubt*
*Every story ends in opening from some future
into past*
Every glory is the story of finding first in last
*Every aching is the making of another role
for play*
*Every taking is the slaking of another's thirst
to stay*
*Every tiding's no confiding with-out the trust
to tell*
*Every siding is no hiding from the fear of
utter Hell*
*Every flowing is the ebbing of another's
world within*

*Every glowing is the lighting of the darkness
in the spin
Every heartbeat is the murmur in the core of
inner space
Every drumbeat is the echo of the dance
within each place
Every silence is the gathering of the storm
that is to come
When Love comes to Life*

Series Editor Preface

Spaces In-Between: The Arena for Development

This book *The Origin of Life Patterns—In the Natural Inclusion of Space in Flux*, written by Alan Rayner, inaugurates the new SpringerBriefs series *Psychology and Cultural Developmental Sciences*. As the firstborn, this book has to accomplish the task of drawing the coordinates for our intellectual enterprise, outlining the theoretical basis and the methodological approach within which a vivid debate at the intersection of cultural psychology and other developmental sciences (biology, sociology, philosophy, anthropology, education, etc.) will be promoted. The aim of this series is to create fertile ground that integrates knowledge from many fields into a new synthesis—a general science of cultural psychology that deals with the highest psychological functions of human beings (Valsiner et al. 2016). Much of contemporary developmental science has natural liaisons with cultural psychology. The SpringerBriefs series creates a forum of scholarly interchanges for that interdisciplinary synthesis.

This book is a programmatic statement focusing on some aspects of the complex phenomenon of becoming human in the social and natural world around us. We have already had the fortune to benefit from the intellectual work of Alan Rayner in 2011, when, on the pages of the Springer journal *IPBS: Integrative Psychological and Behavioral Science* (of which, this series is also a topical extension), appeared a quite unique theoretical elaboration of the notion of inclusionality and the role of the 'space in-between' and borders in developmental processes (Marsico 2011; Rayner 2011). However, much of that space remained uncharted back then—and is accomplished now in the present book.

Rayner's new scientific biological paradigm, here extensively developed, is based on the shift from an abstract passive conceptualization of space as 'void background' to a natural, relational view of space as a 'receptive omnipresence.' This model greatly contributed to our intellectual attempt to elaborate the general notion of the liminality of the human condition. Let us highlight a part from Rayner's conclusion which we consider to be revolutionary for our contemporary social sciences:

> Abstract thinking removes the middle ground of self-identity as a dynamic inclusion of neighbourhood. This explains why two incompatible kinds of abstract logic have been at odds with one another for millennia. 'Two-value logic' (also known as the Law of the Excluded Middle) straightforwardly regards one or other of the two mutually exclusive alternatives (bounded or unbounded) to be 'true' and the other as 'false'. Dialectic logic holds both alternatives to be equally true, which results in paradox. *Since there is no way to resolve this paradox naturally, by allowing boundaries to fluidize and space to be continuous, the brutality of one and the softness of the other are held in 'living contradiction'* (p. 105, emphasis added)

This 'living contradiction' of the two systems of logics has been around for more than two centuries in contemporary European thought, and the 'paradox' of the dialectic still hinders our view of the many 'spaces in between' which Rayner so beautifully and artistically demonstrates in this book on the basis of phenomena from nature, and amply illustrates by his own paintings. Rayner's is a new version of *Naturphilosophie*—attempted two centuries after its earlier romantic versions were attempted (and abandoned). We hope that the readers of this book will pick up the challenge—to give explicit form to the 'fluid and continuous' boundaries that are prominent in nature and in the human psyche. Advancement of new models of developmental mereotopology could be one of the solutions (Marsico 2011). Development of the philosophical principles of 'double negation' (elaborated by Engelsted 2017) could be another.

It is nice to encounter humble authors in our editorial efforts. Alan Rayner did not even realize how central his evolutionary standpoint has been for the psychological investigation over the past several years. From his inspiring work, so many questions about the mutual adaptation between the space and the living system from a cultural psychology perspective have been raised. That is why this book series, which is meant to be interdisciplinary in its nature, could not have any other inaugural book than *The Origin of Life Patterns—In the Natural Inclusion of Space in Flux.*

Alan Rayner is also a brilliant example of how intellectual work is an unceasing enterprise. Over the last months during the preparation of his manuscript, we had intense correspondence that was of great help in understanding the theoretical roots and the methodological options, as well as the integration of science, psychology and arts that the author provides into the volume. It is always fascinating to see a brilliant mind at work! In one of his recent messages, Alan Rayner incisively defined the central theme of this book which seems to be the *compendium* of what he has been elaborating over decades. It has to do with what *The* Nature is, and then, what is the quintessential aspect of *Human* Nature. At the core of Rayner's argumentation there is the idea that in our contemporary scientific, educational, theological and governmental structures, the predominance of a fragmented way of thinking that separates what is material from what is immaterial, and that which is to be considered the 'subjective' and the 'objective' as mutually exclusive, or collapsed one into another, make it impossible to understand the complexity of the human nature in a manner that preserves the wholes while studying carefully their parts. Understanding emerging, developing and self-maintaining wholes is the task of new science.

It is this that the current rapidly developing perspectives, in what are subsumed under 'Cultural Psychologies' (Valsiner 2014), are oriented to do: Keep the focus on the wholes while recognizing the high variability of their constituent parts. Both biological and social sciences operate on phenomena that are characterized by *variability amplification*, as pointed out by Magoroh Maruyama in his crucial introduction of the notion of 'second cybernetics' more than a half century ago (Maruyama 1963). Biological and social systems—open in their relationships with the environment—constantly produce innovation. New forms come into being, which are transformed into still newer forms—while maintaining *generative continuity* with the past. This leads to a number of deep changes in the ways in which scientists need to think about the natural and social orders—moving from thinking in terms of causality to that in terms of catalysis (Cabell and Valsiner 2014), and to the corresponding abandonment of thinking in terms of 'independent' and 'dependent' variables (Valsiner and Brinkmann 2016).

Alan Rayner's book will surely foster reflection on what 'Our True Nature' is— guiding our attention to the way in which any living system is constantly in dialogue with its natural neighbourhood, on the basis of an interdependent and co-evolutionary process involving both the context and the organism. We are sure that *The Origin of Life Patterns—In the Natural Inclusion of Space in Flux* will have a beneficial effect on those who are trying to overcome the traditional borders between basic sciences and humanities (*Geistewissenchaften*). Our ways of knowing are similar across the artificial divide of the two kinds of sciences, created in the nineteenth century. It is time to restore the understanding of that similarity.

Aalborg, Denmark   Giuseppina Marsico
January 2017   Jaan Valsiner

References

Cabell, K. R., & Valsiner, J. (Eds.). (2014). *The catalyzing mind.* New York: Springer.
Engelsted, N. (2017). *Catching up with Aristotle. A journey in quest for general psychology.* New York: Springer.
Marsico, G. (2011). The "Non-cuttable" space in between: Context, boundaries and their natural fluidity. *IPBS: Integrative Psychological and Behavioral Science, 45*(2), 185–193. doi:10.1007/s12124-011-9164-9
Maruyama, M. (1963). The second cybernetics: Deviation-amplifying mutual causal processes. *American Scientist, 51,* 164–179.
Rayner, A. D. (2011). Space cannot be cut: Why self-identity naturally includes neighbourhood. *Integrative Psychological and Behavioural Science, 45*(2), 161–184.

Valsiner, J. (2014). *An invitation to cultural psychology*. London, UK: Sage.

Valsiner, J., & Brinkmann, S. (2016). Beyond the "variables": Developing metalanguage in psychology. In S. H. Klempe & R. Smith (Eds.), *Centrality of history for theory construction in psychology* (pp. 75–90). New York: Springer.

Valsiner, J., Marsico, G., Chaudhary, N., Sato, T., & Dazzani, V. (Eds.) (2016). *Psychology as a science of human being: The Yokohama manifesto* (Vol. 13 in Annals of Theoretical Psychology). New York: Springer.

Acknowledgements

A great many people have contributed, wittingly or unwittingly, to the writing of this book. Given the inseparability of my personal identity from my natural neighbourhood, it could not have been otherwise. It has not been an easy road to travel given the contrast between the way I have come to regard the natural world and the prevailing worldview of modern human culture. I therefore especially want to thank two small groups of people. First my close family, especially my wife, Marion and daughters, Hazel and Pippa, who have witnessed first-hand just how stressful and painful I have found it, and been there to support me lovingly, nonetheless. Then there is that small group of philosophical companions who have stayed alongside, guiding and encouraging me for years on end, despite and perhaps because of my frequent expressions of self-doubt. Finally, there are two correspondents, Doug Marman and Roy Reynolds, whose close attention to a first draft enabled me to augment and refine the final version.

Contents

About the Author

Dr. Alan Rayner is currently President of the Bath Natural History Society. He is an evolutionary ecologist, writer and artist. Dr. Rayner has published in numerous papers and books, including, most recently, 'NaturesScope'. He is a former President of the British Mycological Society (in 1998). Since 2000, Dr. Rayner has been pioneering awareness of 'natural inclusion', the mutual inclusion of energetic flux and spatial stillness in all locally distinguishable phenomena. This enables us to understand ourselves and others as dynamic inclusions of our natural neighbourhood, not independent objects. His special interest is in helping people to become more aware of the diversity of wildlife in their local neighbourhood, and how this can help us to live together in a more passionate, compassionate and sustainable way than we currently do.

Chapter 1
Introduction: A Personal Experience of Culture Shock

Abstract The author's early life experiences that contributed to his appreciation of his own and others' self-identities as receptive centres of awareness are briefly described. This appreciation is contradicted by objective perceptions of reality that exclude individual receptivity from consideration by mentally isolating natural bodies within definitive boundary limits. While these perceptions seem to render life mechanistically predictable and have become deeply culturally embedded, they are falsely premised and are a source of profound misunderstanding and psychological, social and environmental harm and conflict. The author aims to provide a remedy for these problems through a deepened, more natural understanding of the origin of life patterns.

1.1 Early Awakenings

Inescapably, the way we develop psychologically as human beings is profoundly influenced by our earliest life experiences in the prevailing climate and culture that we are born into. Those experiences set the scene for what is to come by way of patterns of perception, thought and behaviour that recur throughout our lives, whether we are consciously aware of them or not and whether we regard them as beneficial or not. Where these patterns have a restrictive influence on our lives, it may be very difficult for us to find a way to break free from them. Where they open our minds to possibilities others are not aware of, they can be a source of great personal creativity and insight, but also great frustration and distress if we cannot communicate our meanings and intentions to those with whom we live most closely.

No doubt, this is why I find myself writing this book now, 66 years after I was born. I am still profoundly affected by what I have increasingly recognised as the contradiction between the perceptions of my place in the natural world that I developed as a young child and the prevailing social and scientific attitudes of mind

© The Author(s) 2017
A. Rayner, *The Origin of Life Patterns*, SpringerBriefs in Psychology and Cultural Developmental Science, DOI 10.1007/978-3-319-54606-3_1

that I encountered as an adult, working as an academic research scientist and university teacher in southern England. I think I can fairly describe my experience of this contradiction as 'culture shock'. This is a kind of trauma that has led me to develop an intense personal interest both in the diverse yet recurring patterns of life found over different scales in natural ecosystems, and in how our perception of those patterns is affected by our personal psychological experiences. I will say immediately that I have come to regard the objectivistic way that most of us have been taught to think in modern culture as profoundly mistaken, rooted in a false premise that contradicts the way we naturally are. By leading us to exclude consideration of our own receptive awareness and thereby view ourselves and others as mechanical products of lifeless matter, I think this false premise leads to serious psychological, social and environmental harm and conflict. I want to show how a deepened awareness of natural patterns of life can enable us to think in the more receptive and simple way that I did as a young child.

So, how did I come to experience this contradiction? First, let me say a little about my early childhood. I was born in Nairobi, Kenya, in 1950. In 1958, my British family (parents and elder sister) returned 'home' suddenly and unexpectedly to live in London, after my father suffered from a stroke. At the time of our departure, my mother had been Deputy Mayor of Nairobi and was expecting to become Mayor the following year. My father was a plant pathologist, working on coffee rust and coffee berry disease. I found myself displaced and entered into the British educational system, in which I struggled badly initially but eventually succeeded in getting to King's College, Cambridge, and embarked on an academic career.

During my years in Kenya, I received rather little schooling, but with the help of my sister I had learned the rudiments of reading, writing and arithmetic. Much of my life was spent playing in a large, semi-wild garden leading down to a muddy river. It was not a very safe place, but I loved especially to climb trees until one day, shortly before our departure, I climbed too high, a branch broke under my weight, and I was very lucky to suffer only a severely broken right arm and gashed leg. But by that time I already had intimate experience of 'wildness'.

One of my most powerful psychological experiences came during one of my mother's political 'sharing circles' in which she invited local politicians of all colours and creeds into our living room to exchange views. Seated with her I remember vividly being struck by the thought that 'I am just as much included in what each of these very different people are seeing as they are included in what I am seeing'. Unwittingly, I was using a combination of 'receptive' and 'discerning' perceptions to appreciate my own and others' place in the world. I recognised that the enormity I sensed within my own body was also sensed within every other body I was looking at, which appeared so diminished and impersonal from a distance. In other words, we all live in each other's common space or 'neighbourhood' and cannot avoid deeply affecting the enormity within one another's lives. That realization led me to develop a strong sense of 'duty of care' for my natural neighbours (human and non-human life forms) and for my natural neighbourhood as well as for myself.

So when I was later told of Jesus of Nazareth's encouragement to 'love your neighbour as you love yourself' it came as no surprise to me. What did come as a shock was when a Christian teacher told me that the Cross symbolizes 'I' crossed out—altruistic self-sacrifice—while I was simultaneously being exhorted by other teachers to compete ferociously with my peers in order to 'succeed' in life. And in due course came other shocks, as when I was told by my Head of Department, during my Ph.D. studies, that I had an 'over-developed social conscience', and was confronted by Darwinian and neo-Darwinian colleagues who believed in 'selfish genes' and 'the preservation of favoured races in the struggle for life'.

So my experience of 'culture shock' followed my displacement from an African world in which I recognised that 'we are all included in the common space of one another's natural neighbourhood', into a world that objectively alienates 'self' from 'habitat' as a matter of inescapable fact and logic. It is a shock that reminds me of the story of 'the Fall'—Adam and Eve's banishment from 'Eden' following upon their temptation to discriminate definitively between 'Light' and 'Darkness' as sources of 'Good' and 'Evil'.

In the early 1970s, I found myself painting two pictures that symbolize the contrast between my world view as a child, and the objectivistic world view that confronted me as an adult (Fig. 1.1 and 1.2)

Fig. 1.1 'Tropical involvement' (Oil painting on board, by Alan Rayner 1972)

Fig. 1.2 'Arid confrontation' (Oil painting on board, by Alan Rayner 1973)

In this book, I combine personal testimony with careful reasoning and scientific knowledge to show why I think a 'Return to Eden' is both intellectually and emotionally urgent, if we are to live wisely, lovingly and sustainably with the scientific knowledge that we have gained of the natural world and ourselves. My writing will mirror the psyche's journey from conception to maturation in a culture not entirely of its own making. You might view it like the course of a river surging its way around, under and over obstacles as it sustains its natural continuity from source to sea. Much of it, so far as I am aware, is without published precedent other than in my own writings. Perhaps the nearest philosophical approach is that of 'phenomenology' (e.g., Merleau-Ponty 1945; Abram 1997). I ask readers to bear this in mind when comparing my personal account with what they hear, read or are taught elsewhere, and to regard what I say as a stimulus for their own free thinking, not as a definitive 'last word'. What I want to do, together with others, is assist into life a new and deepened receptive awareness, understanding and appreciation of the creative diversity of Nature and our human inclusion within it.

References

Abram, D. (1997). *The Spell of the sensuous*. New York: Vantage Books.
Merleau-Ponty, M. (1945). *Phenomenology of perception*. London: Routledge.

Chapter 2
Noticing Recurrent Patterns, from Microcosm to Macrocosm

Abstract Careful study reveals a variety of recurrent natural patterns of life that are evident over vastly different scales of organization, within living cells, multicellular organisms, colonies, populations, communities and ecosystems: stars and stripes, circles, lines, spirals, rivers, ripples and crazy paving. The vital role of receptive perception and imagination in the personal recognition and appreciation of these patterns is emphasized. The suppression of these qualities by the imposition of rationalistic logic and language that objectively isolates observer from observation is seen to be a source of profound misunderstanding, which detracts from the quality of human life experience, learning and discovery. By contrast, the value is recognized and exemplified of a discerning approach to natural scientific enquiry that incorporates and encourages empathic personal observation and feeling awareness rather than relying completely on the abstract findings, methodologies, theories and truth claims of prior and current authorities. Such an approach to enquiry, which reciprocally combines receptivity and discernment, rather than divorcing them as abstract 'subjectivity' and 'objectivity', is available to anyone, not just those privileged by a formal scientific education. Most fundamentally it requires *imagination*, awareness of intangible presence and possibility, which is beyond the comprehension of objective rationalization. Its exposition may hence include art and poetry, as well as analytical discourse.

2.1 Evolutionary Ecology: The Art and Science of Recognizing and Understanding Pattern, Process and Relationship in the Natural World

My overall aim in writing this book is to identify the recurrent patterns in which life is expressed over diverse scales in natural ecosystems and to explore how a new awareness of their evolutionary origin in the natural inclusion of space in flux can be related to human cultural and developmental psychology. I will explain why these patterns cannot adequately be represented or understood in terms of abstract logic and language that definitively dissociates the material content from the spatial

© The Author(s) 2017
A. Rayner, *The Origin of Life Patterns*, SpringerBriefs in Psychology and Cultural Developmental Science, DOI 10.1007/978-3-319-54606-3_2

context of natural systems. Both biological and psychological patterns can, however, readily be appreciated as expressions of natural energy flow and how this both
influences and is influenced by mechanical and mental resistance to movement.

Correspondingly, my core theme will concern how the perception of natural
space as an infinite, intangible, receptive presence, and of natural informational
boundaries as continuous energetic flux, revolutionizes our understanding of evolutionary processes. The *mutual natural inclusion of receptive space and informative flux in all distinguishable local phenomena* enables evolutionary
diversification to be understood as a fluid-dynamic exploration of renewing possibility, not an eliminative 'survival of the fittest'. Self-identity is recognized to be a
dynamic inclusion of natural neighbourhood, not a definitive exception from
neighbourhood (Rayner 2011a, b).

In mathematical and scientific terms, I will show how abstract Euclidean,
non-Euclidean and fractal geometry can be transformed into natural flow geometry
through the incorporation of receptive space and intrinsic flux into mathematical
figures as flow forms. I will describe how the fundamental principle that energy
travels in pulses and resides in circulation applies to all organizational scales, from
microcosm to macrocosm. Finally, I will reveal how the discontinuous perceptions
of space and boundaries that are deeply embedded in many cultural and scientific
traditions have contributed to the emergence of abusive power-relations, psychological distress, environmental damage, and social conflict. I hope thereby to offer a
new approach to philosophical enquiry that can enable us to remedy these human
problems.

And so, let us commence, with an introduction to my personal approach to
receptive enquiry and how this can be used to reveal and gain insights into the
recurrent patterns of life that I will be seeking to account for in subsequent chapters.

2.2 Receptive Enquiry: Exploring Natural Patterns with Eyes and Mind Wide Open

As an experienced and enthusiastic naturalist, I often lead, or co-lead, 'nature
walks' for members of the public. These can take place anywhere—not just what
are sometimes called 'sites of special scientific interest'. A local park, a residential
street, a bit of woodland, a grassy sward—any or all of these would be suitable. It is
not unusual for participants to say something like: 'I've walked past that every day
without noticing or appreciating it'. And yet many of those same people will spend
a significant proportion of their earnings, and consume substantial amounts of fossil
fuel, travelling to somewhere exotic in search of 'new experience'. What kind of
commentary on modern life and education is this? How come we have fallen so far
out of touch with our natural neighbourhood?

Over the years of my life, and in the light of my own personal experience of
academic education, I have concluded that for deep-rooted reasons, generation after

generation of adult human beings has been prone to teach its offspring to neglect its natural neighbourhood. This is psychologically, socially and environmentally damaging. While computer technology and the Internet have opened up vast learning opportunities, the way they are mostly used has aggravated rather than alleviated our sense of alienation from the world we inhabit. We need instead to start paying attention to what we live our lives amidst in a way that opens us up to the deep and varied lessons that we can learn from it. We need to recover our childhood and aboriginal sense of wonder in what naturally surrounds and includes us. We need to explore our environment with eyes and minds wide open to the insights and understandings it offers concerning our natural origins and needs, instead of imposing our own abstract preconceptions and rules of conduct upon it. And, as I have previously discussed (Rayner 2011a) and will try now to exemplify, we need to use evocative language and imagery to express our findings in a lively, sensuous, appreciative way.

Never Quite Knowing

Life is a creative exploration of renewing possibility,

Not a competitive struggle for permanent existence –

Poetry, not Prose

Improvisation, not Prescription,

Tolerance, not Rigidity,

A Search for Openings, not Quest for Completion

.

Motion in Stillness, Stillness in Motion,

Responsiveness in Receptivity, Receptivity in Responsiveness,

Energy in Space, Space in Energy,

Not One or Other Alone,

No matter without no matter

.

Never Quite Knowing

What's coming next,

Preparing for Surprise,

Ready to change One's mind,

One's direction

.

That's the evolutionary learning curve

In natural inclusion –

Truly natural Science,

Truly natural Art

Exploring natural neighbourhood with Love

Exciting and Inspiring

Isn't It?

(composed on 22nd October 2015, with appreciation of Emily Dickinson, 'I Dwell in Possibility').

The sheer joy of encountering all the diverse and wonderful patterns in which life expresses itself in the natural world, and recognizing how these patterns recur at so many different scales and in so many different situations, has always been my principal reason for loving natural history. This has been a far greater source of inspiration for me than the simple hunter's quest to find and name as many different species of wildlife as I can. Or, rather, it is why I enjoy this quest so much: because becoming good at it requires me to learn to recognize what is known as 'jizz'— recurrent patterns of appearance as basic underlying 'themes' expressed in myriad 'variations'. This same intuitive ability is what enables us all to come to recognize a familiar human face, regardless of its expression and hairstyle and without need to pause for analytical thought.

It is the recognition of those underlying themes in Nature that brings me the most entrancing and meaningful pleasure—a way into understanding how and why these recurrent patterns come into being, which takes me very deep into the fundamental questions that contemplative human beings have always pondered. My love of natural patterns led me to uncover the role of what I nowadays call 'natural inclusion' as the universal underlying relationship from which all patterns of life—and not just 'biological life'—ultimately arise. In the process of this 'uncovering', I had to work my way down imaginatively through many veils of superficial 'appearance' to arrive at the simple relationship between movement and stillness from which those appearances emerge. Here, I invite you to accompany me down through those layers, and wonder if we will find ourselves arriving in the same place.

So let me now take you on a 'nature walk' with me. As we explore receptively, with eyes and minds wide open, let me ask *you*, as an appetizer, what recurrent themes *you* notice as you explore the natural world. Do you see stars and stripes, circles, lines, spirals, ripples, rivers and crazy paving? If so, where? Do you see any other kinds of recurrent pattern that I have not mentioned?

Imagine yourself included in this scene (Fig. 2.1).

Has this scene always been as you see it here, and will it always remain as you see it now? If not, how has it come into being, and how will it change? What makes the forms you see distinct from one another and their surroundings? Why is it that you do not just see a featureless monotone, but instead apprehend a marvellous array of shapes, shades and colours? Why, with yourself included amongst what you see, are these visual distinctions added to by a rich variety of sounds, smells, temperatures and soft and hard textures, including those within your own body—all of which combine into your sensation of being miraculously present as an *inhabitant* of the scene, throbbing with life?

Fig. 2.1 Members of Bath Natural History Society taking lunch during a meeting at Newark Park, Gloucestershire (photograph by Marion Rayner)

Now, while still imaginatively including yourself and your feelings as an observer within the scene you are observing, not standing aloof from it, let us pay a visit to one of the inhabitants of the scene, for example a tree. From a distance we might just make out the appearance of something like a fuzzy lollipop, sticking out of the ground. As we move in closer, more and more of its branching structure becomes discernible, shooting out from its trunk into smaller and smaller offshoots, culminating in buds, leaves and flowers. As we move beneath its canopy, we become aware of the shade it casts over us and the diversity of life forms that reside beneath its cover. We notice the airiness that is all around and amongst its branches and foliage, sometimes quite still, other times whispering gently or even rushing stormily. If we look closely enough, with the aid of a magnifying lens, we can witness where this airiness enters the tree's leaves and corky covering, through pores known technically as stomata and lenticels.

Meanwhile, we can sense the varying smoothness and roughness of the tree's bark. We might hear the calls and fluttering movements of birds, and catch sight of their diversely coloured and patterned plumage in flight or perching. A squirrel may distract us. Examining some tree leaves closely, we might encounter some insect or arachnid life finding a home or source of food amidst the ribs and networks of veins (Figs. 2.2 and 2.3).

Fig. 2.2 Close-up view of the spider, *Paidiscura pallens*, and its spiky egg-case, on the underside of an oak leaf (photograph By Rob Randall)

Fig. 2.3 Larva of green silver-lines moth, *Pseudoips prasinana*, with patterning of lines and spots, on underside of a hazel leaf (photograph by Suk Trippier)

Surveying the tree's trunk, branches and the ground into which its roots spread, we might notice the wonderful rippling, branching, sprouting, erupting, collapsing variety of lichens, mosses, liverworts and vascular plants that find a home here (Figs. 2.4, 2.5, 2.6, 2.7 and 2.8).

So, too might we notice the fanning, mushrooming and splurging fruit bodies of fungi that find food and shelter deep within the tree's woody interior and decomposing remains, as well as those forming what are known as mycorrhizal partnerships with the tree's roots. If we were to dive down microscopically with the tubular hyphae of those fungi into the wood of the tree, we would discover it to be full of holes—the communicating pipelines that link water-absorbing roots with photosynthesizing canopy—not the solid block of substance, it might seem to be at first (Fig. 2.9).

Now, coming out from the shade of the tree into the bright light of day, our attention might switch to wider vistas, to notice the form of the landscape with its hills and valleys rising skywards and descending towards sea level. We might notice the branching, sinuous watercourses that run through the valleys, simultaneously shaping and being shaped by their banks in a mutual correspondence of each with the other. We might notice flocks or herds of grazing animals either scattered around the hillside or following one another's footsteps from one pasture to another, along worn-down paths that resemble the convergences and divergences of the watercourses. Lifting our eyes skywards, we notice birds in flight, either as singletons or in pairs or groups with few to thousands of members. Amongst the latter may be what are known as 'murmurations' of starlings, co-creating smoke-like patterns as they twist and turn together in dynamic relationship with those air currents that carry shape-shifting clouds along with them. Where the sky is clear we may catch painful sight of the sun's orb, or notice the pallid disc of a daytime moon.

Notice how our attention has shifted with this move out into the open, from close-up and intimate to far away. While exploring under the tree's canopy we felt enveloped within and by the life abounding there. Outside it, standing upright and gazing from a fixed *viewpoint*, the vision provided by the binocular eyesight characteristic of primate mammals (i.e. with eyes on the front of our faces) stretches in classical perspective towards the horizon where all that seems large in foreground shrinks to *vanishing point*. We feel a static sense of remoteness from what we are observing horizontally. We experience a sense of elevation over the *substance* beneath our feet, as if the self within our bodies is outside or above it all. It is a powerful illusion that can very easily go to our heads, alienating us from our surroundings. This is the illusion that makes us perceive and analyse the natural world about us in an abstract, photographic way. It is as though a shift takes place in our perception of space as a receptive presence everywhere in which we feel ourselves immersed like fish in an ocean, to a purely external background void from which we feel isolated. Mcluhan (1967) alluded to this as a shift from 'acoustic space' to 'visual space', and it is at the root of abstract, objectivistic scientific thought and method. We place the world within a fixed frame of space and time that does not include us, and incorporate this distanced perception into our definitive

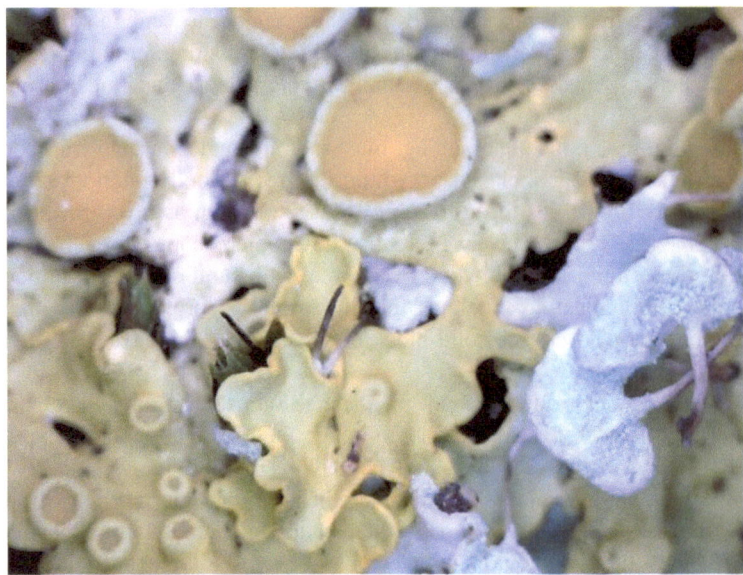

Fig. 2.4 Close-up view of the lichens, *Xanthoria parietina* and *Physcia adscendens* on a tree branch

Fig. 2.5 Close-up view of Bruch's Pincushion moss, *Ulota bruchii*, growing on a tree branch (photograph by Paul Wilkins)

Fig. 2.6 Rippled fronds of Hart's-tongue fern, *Phyllitis scolopendrium*, with stripe-like spore-producing sori on their undersides, growing on shady bank amongst trailing ivy stems and foliage (photograph by Marion Rayner)

Fig. 2.7 Unfurling spiral of a fern frond (photograph by Marion Rayner)

Fig. 2.8 Shoots and starry flowers of Wild Strawberry, *Fragaria vesca* (photograph by Marion Rayner)

Fig. 2.9 Fruit bodies of Turkeytail fungus, *Trametes versicolor*, with multi-coloured, concentric, glossy and furry banding, growing out between the cracked bark of a decaying tree branch (photograph by Marion Rayner)

logic, language and belief systems. We may even come to regard and analyse ourselves in the same way, as remote objects that we imaginatively stand outside and scrutinize.

All it takes to overcome this *objective* distancing is to 'come back down to Earth', bringing what we are observing back into the intimate, sensuous range within which we feel the enormity within ourselves immersed. We might get down onto our hands and knees and scrutinize the miniscule life forms living amongst the grass and soil. Or we might spin our bodies around, to take in the panoramic, self-including view that comes naturally to grazing animals, with eyes on the sides of their heads instead of the front of their faces.

The Humility of the Valley

Life doesn't strive

To secure its foundation

Upon the rocky serrations of the High-minded

Where Men build castles in the air

To furnish that false sense of superiority

Which comes from the pretence

Of overlooking all around

To the edge of infinity

.

Life thrives

In the seclusion of the valleys

Where dampness accumulates

In the earthy humidity

Of humility

Warmly tucked in

To the bed of sea and land

Rich with variety

Exuding

Intruding

Out and into the cosiness

Of each lovingly enveloped

In the other's influence

.

Wisdom cannot be found

On peaks of adaptive fitness

Running with Red Queens

But only in that radiant depth

That reaches everywhere

Through the heart of somewhere

(Rayner 2011b).

There is also another way in which our perceptions of the world about us can change radically, and that is to experience it at night—an experience that the bright lights of the city have rendered strangely unfamiliar for a great many modern-day human beings. And it is an experience that time immemorial has brought a powerful combination of fear and awe to human beings. The fear is due to the uncertainty that comes when we lose our ability to see where we are. Without this ability, we cannot detect potential dangers or find what we are looking for. We feel vulnerable. We crave the certainty that our eyesight affords, and so yearn for light to banish darkness. Such is the reliance of sighted people on our visual sense that we may even come to associate darkness with 'evil' and death. But in trying to gain certainty by excluding ourselves from darkness's reach, not only can we also lose our sense of awe but we can also reinforce the visual illusion that distances us from our natural neighbourhood, and even from ourselves.

Let us now imagine ourselves on that hillside in Fig. 2.1, as sun sets and dusk sets in. The first thing we may notice as the clarity of our vision begins to fade is an increased sensitivity in our hearing, which picks out bird and animal calls. It also picks out the silence within and between and within those punctuations, which turns them into music. Landscape becomes soundscape. Nocturnal animals, whose active time of day is night, start to make their presence known. We may catch a glimpse of the flickering wing beats of bats and moths, the ghostly, silent fluttering of a barn owl or hear the hoots of a tawny owl. Slowly darkness enshrouds us and we begin to rely on heightened tactile senses to grope and feel our way around. The darkness itself is not painful, though the bites of midges homing in on our warm bodies may be. Then as our night vision begins to adjust, and providing there is a lack of cloud cover, we notice the stars coming out and the pallid disc of the daytime moon transforming into shimmering brilliance. The darkness does not prevent the light from these heavenly bodies from reaching us, but instead offers it free passage. For, this kind of darkness is simply the transparency of space, and not light-absorbent fabric. The frictionless, transparent, silent darkness of space is not an enemy of light. It is a natural presence that permeates everywhere, without restriction but only becomes noticeable when our visual senses are not flooded with illumination—as also occurs during total solar eclipses. Nor is this darkness 'cold'—in reality it is a thermal insulator, as we make use of in vacuum flasks that keep their contents cold or hot. This darkness is not in itself anything to be fearful of—it is only what our imaginations make of it that can be truly terrifying.

You, Darkness

You, darkness, that I come from

I love you more than all the fires

that fence in the world,

for the fire makes a circle of light for everyone

and then no one outside learns of you.

But the darkness pulls in everything-

shapes and fires, animals and myself,

how easily it gathers them! -

powers and people-

and it is possible a great presence is moving near me.

I have faith in nights.

Rainer Maria Rilke.

The darkness of space is not, however, the only natural presence that can escape our notice or terrify us as we terrestrial primates go about our daily affairs. And while, as subsequent chapters will show, the darkness of space is the birthing place for matter itself, this other natural presence is the birthing place for organic (carbon-based) life itself, as we know it here on planet Earth. It is the presence that accounts on average for around 65% of the weight of a human body and 99% of its molecules. We nonetheless fear it, especially when it is gathered in large quantities, as a source of great uncertainty and hidden depth within which we, as air-breathing creatures trying to stay afloat can all-too-easily become exhausted, lose body heat and drown. It is, of course, water.

I must go down to the seas again, to the lonely sea and the sky...

For the call of the running tide,

Is a wild call and a clear call that may not be denied...

John Masefield.

If we were to dive down beneath the surface of John Masefield's land-dwellers' impression of the open sea, we would find a place that is often anything but lonely. Alternatively, as the running tide draws its liquidity back from high to low, an astonishing scene is revealed to those of us prepared to open our senses to what lives there. So let us now extend our walk down to that dynamic interfacing where sea meets land in mutual encroachment. Here is a painting, entitled 'Intertidal Highlands' that I made some years ago, based on observations made on the rocky seashores of Gower peninsula in South Wales (Fig. 2.10).

Here, laid out before us in myriad guises are examples of every kind of life pattern that I will consider on this book. Organic life on Earth is widely thought first to have evolved in the sea, and it is here that the capacity for living pattern generation in all its fluid-dynamic possibility first flourished. Emergence of life from this aqueous nursery onto land entailed exposure to high light intensity, wind, desiccation, rapid diffusion of gases such as oxygen and carbon dioxide, while

Fig. 2.10 'Intertidal highlands' (oil painting on board by Alan Rayner 1998)

leaving behind the buoyant support and powerful currents of water as an external as well as internal milieu. The effects of these environmental changes on bodily form are evident in the distinctive zonation patterns found below, above and between low and high tide marks. I often reflect that our view of life and neighbourhood would be very different—and less prone to objective abstraction—if we were truly marine creatures rather than terrestrial apes predisposed to single out and grasp what we need to feed, clothe and shelter ourselves. Here is a painting and a poem based on that reflection, called 'Landed Stranded' (Fig. 2.11).

Landed, Stranded

A reflection upon the evolutionary inversion from aquatic to terrestrial life

I used to be

Within the Sea

An identity

Of You and Me

Submerged

In Commonality

Of Sounding

Between Airy Heights

Fig. 2.11 'Landed stranded' (oil painting on canvas by Alan Rayner 2004)

And Bottom Depths

Waving Correspondence

Through Inseparable Togetherness

Of Content with Context

.

But, Now,

Dry

Abstracted

Space comes between Us

A separating distance

An unbecoming Outside

Alienating Forms

As Fixtures

Stranded in Isolation

Entities

Non-identities

Conflicting

Oblivious of Our Belonging

Together

.

Oxygen

Now, moving Fast

Not Languidly

Tans our Hides

Protecting Our Inner Space

Against its own

Consuming Presence

Supporting Combustion

Burning Us Out

.

But all this sealing

Removes Our Feeling

Setting Our Content

At Odds with Our Context

So that we push

Against the Pull

With Backs to Front

Itching to Relieve

Unbearable Friction

.

And So Now

Just Let's Go

And, with Loving Fear

Dive into the Clear

And Swim Where it's Cool

To be In With the Pool

Together.

Fig. 2.12 'Recalcitrance' (oil painting on canvas by Alan Rayner 2001)

So far, the examples I have chosen have all been what are known as 'macroscopic' forms of biological life, because we can readily observe them with the naked eye or a low-power magnifying lens. The recognition that these forms are actually composed of what are typically microscopic structures called 'cells', was undoubtedly one of the most significant discoveries of modern science. It set the scene for all we now know about genetic inheritance and how the biochemistry of life is organized within fluid systems of membranes and organelles. My painting above represents how the outward life of a 'Star Thistle' plant (*Centaurea calcitrapa*) arises from the inner life of a vulnerable body of cells living and functioning collectively (Fig. 2.12).

It is perhaps unfortunate, however, that on account of their immediate appearance when examined under a microscope, cells were ever actually called 'cells'. To do so reinforces the misleading impression of them and their molecular components as the rigidly self-contained units and sub-units that abstract science and mathematics has been all-too-ready to depict popularly as 'building blocks of life'. In reality, cells are not assembled from building blocks, nor are they assembled into multicellular bodies by some mysterious building agency. They are, like all the forms of life that they arise from and give rise to, fluid-dynamic localities bounded by a membrane that sustains an electrical potential difference between its inner and outer surfaces. This potential difference arises from the use of chemical energy ultimately derived from sunlight or geothermal or chemical sources. In other words, cells and their contents are dynamic natural inclusions of their neighbourhood, not independent objects. No sooner is their potential difference discharged, for example by the ice-damage portrayed in my painting, then they die and the energy they contain is released for take up elsewhere. When present together in multicellular bodies, which arise from the proliferation of parental cells (as when an egg cell divides following fertilization to form an embryo), they depend on communication with one another through their enveloping membranes and (where present) 'cell walls' in order to function coherently.

Are you seeing what I am seeing? Those same recurrent patterns, wherever you care to look for them, at whatever scale from tiny to enormous, both within and far beyond the bodies of Earth's living systems: stars and stripes, circles, lines, spirals, ripples, rivers and crazy paving—plus a myriad variations and combinations of these basic themes.

Would not it be good to be able to understand how and why these patterns arise? I have certainly always wanted to be able to do so. Being an enthusiastic natural scientist, it was to conventional evolutionary theory and mathematics that I first looked for an answer. But after much researching and soul-searching, I realized I would *never* find an adequate explanation this way, because the underlying objectivistic framing of those approaches actually depends on excluding from consideration those very presences throughout Nature that are so vital to the very possibility for patterns to form.

Instead I literally drew insight from my own efforts to illustrate diverse patterns of life in paintings, such as this one, 'Future Present', in which I sought to feature every major group of life forms currently resident on planet Earth and their watery origin as expressions of genetic code (Fig. 2.13).

If you want to get a *feeling* for what you can observe in the natural world, I think there is no better way than actually going through the physical process of trying to draw, paint or sculpt it with hand, eye and brain working together. Instantaneously capturing it in a photograph, or letting a digital computer program figure it out for you will not work, and—for reasons I will describe shortly—can even be profoundly misleading. Try it for yourself. *Honestly* try to draw or paint a flower *as it actually is*, for example.

It was quite obvious to me that I could only bring my painting of 'Future Present' into existence by combining stiff paint pigment with loosening solvent on a receptive surface and moving it around. Paint pigment alone would remain just a blob. I needed to add solvent and then *work* it into the variety of shapes, shades and colours of the painting.

What does this imply about the fundamental creative process through which any form or pattern comes into being? The simple answer I explore in this book is this

Natural creativity is due to the inclusive relationship between emptiness, movement and bodily resistance to movement, i.e. space, energy and material inertia.

Fig. 2.13 'Future present' (oil painting on canvas by Alan Rayner 2000)

For any form to come into being, there have to be at least two basic kinds of presence in Nature: a motionless, *receptive* presence (i.e. a form-receiving presence, analogous to empty canvas) and an *intrinsically* mobile, *informative* presence (i.e. a form-giving presence, analogous to fluid paint). Moreover, these two kinds of presence must *include each other*: the receptive presence alone would be vacant—analogous to empty canvas—and the informative presence alone would have nowhere in which to move around and so give rise to forms of diverse shapes and sizes. In the most fundamental and universal sense, applicable literally to *everywhere*, these two natural presences are experienced in the feeling of emptiness that we call 'space' (what Mcluhan called 'acoustic space', see above) and enlivening flux that we call 'energy'. Energy paints the variety of the natural world on an intangible canvas of receptive space.

Quite simply

- Receptive space is a naturally occurring emptiness everywhere that is not a substance but *invites energy to inhabit it with substance*. It is motionless, frictionless and hence provides freedom for movement. Whether viewing from outside-in or inside-out the distinguishing boundary of any natural form, it is impossible to reach a *limit* where space ceases to exist. Hence it is a truly *infinite* presence
- Energy is continuous, natural motion. It comes packaged into local substance as 'matter', and disperses as 'radiation'

I will elaborate on this central theme of **natural inclusion** *as the mutual inclusion of* **receptive space** *and* **informative flux** *in all distinguishable natural occurrences* throughout this book. Energy and space combine into local material bodies as *flow forms*, which, while being made from movement, resist being moved or deformed out of place to varying degrees depending on their composition and circumstances. These flow forms can be thought of as 'energised beings', whose association, dissociation and exchanges of energy with and from one another is the source of evolutionary variety across all scales of natural organization, including biological organization.

I recognize that this is a very unfamiliar way of understanding the relationship between natural space, energy and matter. To my knowledge it has not been published *explicitly* by any previous author, because of the way these presences have been construed by abstract thought as derivative quantities of 'separation', 'force' and 'mass'. But when we think about it this way, every natural bodily form is actually 100% space plus energy, not part space and part mass. Energy is what brings life to space and space is what enables energy to be expressed in a vast variety of flow forms that are neither completely definable as 'particles' nor uniformly spread out as 'fields'.

That is why conventional analytical science could not help me to understand pattern formation and recurrence—because it is based on *excluding* space, as void background, from energy and is hence incapable of comprehending their mutual, *co-creative* relationship in material bodies. In effect, it *isolates* 'paint' from 'canvas' and so arrests energy within a mathematical point of material with no size or shape, while rendering space into a passive absence of presence, without receptive influence.

As I will expand upon in Chap. 3, conventional science and mathematics seek objectively to *capture* the infinite receptive expanse of space within a three-dimensional frame, and to construct all forms from dimensionless points into breadth-less lines, depthless planes and solid figures. *It does not work*. Both the receptivity of space as a frictionless presence and the intrinsic mobility of energy as an informative presence are omitted from consideration, leaving us with a para-doxical set of solid building blocks from which nothing in reality can be brought into existence! How many schoolchildren, I wonder, have been puzzled by this practice? Even Einstein abstracted matter/energy from space/time, a mental sever-ance that has continued to undermine his and others' efforts to reconcile relativity theory with quantum mechanics.

References

Mcluhan, M. (1967). *The medium is the message: An inventory of effects* with Quentin Fiore, produced by Jerome Agel; 1st edn. Random House; reissued by Gingko Press, 2001.

Rayner, A. D. (2011a). Space cannot be cut: Why self-identity naturally includes neighbourhood. *Integrative Psychological and Behavioural Science, 45*, 161–184.

Rayner, A. D. M. (2011b). *NaturesScope: Unlocking our natural empathy and creativity—An inspiring new way of relating to our natural origins and one another through natural inclusion.* O Books.

Chapter 3
Natural Flow Geometry: 'Pulse' and 'Circulation'

Abstract Abstract geometries, whether Euclidean, non-Euclidean or fractal, cannot realistically describe the dynamic nature of life patterns. This is because they all depend on the unnatural dissociation of matter from space, as if matter and space are independent from each other. This leads to either a reduction of particles into dimensionless point masses or integration of distinctive forms into seamless whole objects. This problem goes away if we use a mathematical approach that incorporates the unlimited emptiness of receptive space and intrinsic informative flux of life, offering more realistic numerical and geometric representations of natural expressions of energy and matter, called 'flow-forms'. The resultant flow-geometry accounts for all discernible natural patterns of life as expressions of the mutually inclusive nature of the empty receptivity of space and informative flux of energy. This natural geometry is referred to as 'place-time', rather than 'space-time'. A basic principle of this geometry is that energy naturally circulates around local, zero-point centres of space, creating residential flow-forms that resist movement out of place, and pulses along linear trajectories from one locality to another, creating communication channels and travelling flow-forms. The inclusive relationship between space, energy and material inertia produces and modifies natural currents, giving rise to myriad variations upon a theme, in the same way that the pooling and flow of water both shapes and is shaped by variably resistive landscape.

3.1 Rigidization by Digitization: The Estranged Square World of Abstract Geometry and Arithmetic

I will preface the first part of this chapter with a warning that even those with a strong mathematical background may find it perplexing, because of the complications and inconsistencies that arise from assuming that matter can be dissociated from space. Setting this assumption aside removes these difficulties. We can then easily see that massy bodies without volume simply cannot and do not exist. The importance of this will be seen in the second part of this chapter, and it may be that some readers would rather take it from there. But, for those who are interested in the

A. Rayner, *The Origin of Life Patterns*, SpringerBriefs in Psychology and Cultural Developmental Science, DOI 10.1007/978-3-319-54606-3_3

historical precedent, it may be rough at times as we come to terms with the deeply habituated objective way of thinking and calculating that has been blocking our view of natural reality for millennia.

Let us begin the rough ride with these statements made by two very famous and respected mathematical scientists

> When a smaller box s is situated, relatively at rest, inside the hollow space of a larger box S, then the hollow space of s is a part of the hollow space of S, and the same 'space', which contains both of them, belongs to each of the boxes. When s is in motion with respect to S, however, the concept is less simple. One is then inclined to think that s encloses always the same space, but a variable part of the space S. It then becomes necessary to apportion to each box its particular space, not thought of as bounded, and to assume that these two spaces are in motion with respect to each other. (Einstein 1954). Space is another framework we impose upon the world ... here the mind may affirm because it lays down its own laws; but let us clearly understand that while these laws are imposed on our science, which otherwise could not exist, they are not imposed on Nature...Euclidian geometry is ... the simplest, ... just as the polynomial of the first degree is simpler than a polynomial of the second degree ... the space revealed to us by our senses is absolutely different from the space of geometry. (Poincaré 1905)

These statements point to a fundamental problem embedded in the foundations of mathematical logic, with regard to its treatment of the relationship between space and material form. There is a clear disparity between treating matter and space as if they are either mutually exclusive from one another or coextensive (one and the same as each other), and our actual bodily experience of life as receptive and responsive inhabitants of a discernible natural world. The former simply cannot truthfully represent the latter.

As I indicated in Chap. 2, the problem is due to the dissociation of observer from observation that comes from viewing our natural neighbourhood objectively, from a fixed standpoint. As soon as we isolate ourselves mentally as a fixed point from what we are viewing, we see what we are viewing as an *instantaneously present* object or set of objects within a static box frame of space and time—as in a photographic perspective. It then becomes conveniently easy arithmetically to divide the 'whole' of this box frame into progressively smaller units, until we reach or impose a 'limit' beyond which it is cannot be divided any further. At that point we are left with a set of indivisible numerical and geometric entities from which all larger entities can be reassembled by means of simple *addition* of one to another. This *analytical* approach is what led to the philosophy of *atomism*, pioneered by Democritus, and it remains with us to this day deeply embedded in the abstract, *materialistic* way we have taught ourselves and our children to *rationalize*, *quantify* and *calculate* using what amounts to a set of rigid 'building blocks'.

Atomism, which is also known as 'reductionism' or 'positivism', brings an illusory sense of certitude to the way we think, leading us to believe in clear demarcations and fully predictive science. But it does so by excluding continuous natural space and motion from consideration by imposing a three-dimensional box frame upon them. Despite its practical utility as a calculating and mapping tool in unchanging or repetitive circumstances, its underlying logic is deeply and profoundly unrealistic when it comes to understanding and accounting for the uncertainty and variability implicit in natural evolutionary processes.

Atomistic thought found one of its most potent forms of expression in the abstract geometry of Euclid, whereby material form is constructed from zero-dimensional points, one-dimensional lines, two-dimensional planes and three-dimensional solids. And it persists also in Peano's arithmetical formulations and in the binary or Boolean logic upon which modern digital computers operate, by splitting the figure, '1', as a point of matter, aside from zero as no matter: 'something or nothing'. Even the Riemannian and Lobachevskian non-Euclidian geometries of convex and concave surfaces originate in this false dichotomy.

The problem with this formulation is that it leaves us with the question of what, then, provides the necessary continuity to assemble points, lines and planes into solid (or even liquid or gaseous) bodies with measurable breadth, length and depth? Since the continuity of natural space as an intangible presence and the intrinsic mobility within matter have been excluded from consideration in this mythical box frame, the only answer available is in the guise of some 'outside-the-box' external causative agency or 'force'. The existence of this agency is then regarded as a matter of mathematical faith that is not to be questioned for fear of unleashing the uncertainty that has so stringently been eliminated from it. This causative agency can even be quantified in terms of its observable effects, as in Newton's Laws of Motion

$$Force = Mass \times Acceleration$$

Meanwhile, the precepts of atomism have increasingly been opposed in recent decades by an alternative, 'holistic' approach to the problem of natural continuity, which seeks to unify or integrate the 'Many' diminutive particulate entities originating from subdivision, into 'One' Seamless Whole. According to 'Gestalt Theory' (Smith 1988), this whole is perceptually present as a 'figure' discrete from its surrounding spatial 'ground' *before* subdivision into parts, not the result of additive assembly from its parts. As such, its original 'wholeness', which is obscured by subdivision, can be restored by holographic means, as in 'magic eye' pictures. This is in some senses akin to integrational calculus, which restores the spatial continuity lost from a natural curve by differentiation into infinitesimal linear sub-units. It leaves unresolved the question of *how* the whole comes into being in the first place, and whether it is truly a bounded identity complete in itself, isolated from its surroundings, or a boundless identity indistinguishable from and hence coextensive with space. Either way, the 'integral whole' is commonly depicted as 'more than the sum of its parts', or, more accurately as 'other than the sum of its parts', but how a seamless entity can have 'parts' is also not explained. Ultimately it is as paradoxical and unrealistic a product of instantaneous, distanced perception as the additive assemblies of dimensionless point-masses that it seeks to subsume.

3.1.1   Statistics, Complexity Theory and Fractal Geometry: Abstract Mathematical Attempts to Account for Natural Uncertainty

The fact that natural patterns of structure and behaviour are actually far more fluid, irregular and less predictable than abstract logic implies has led to two mutually contradictory approaches to incorporating uncertainty into definitive mathematical models of reality. Both implicitly incorporate a role for intangible space, but in radically different ways.

The first, and for a very long while the only approach corresponds with atomistic philosophy and effectively treats space as a source of discontinuity between independent or 'random' occurrences. This is the statistical or stochastic approach. The second, holistic approach came to the fore in the late 1960s and in effect treats space as a domain for internal differentiation within a collective 'whole'. This is the deterministic approach of nonlinear dynamical systems theory and fractal geometry which are sometimes combined under the general heading of 'complexity theory'.

According to the statistical approach, the default condition of nature in the absence of any constraining or ordering influence is *pure randomness* or *maximum entropy*, a set of independent occurrences. Treating space and time as separators so thoroughly disrupts natural structure as to render it into an infinite number of particulate singularities in a sea of emptiness. For any kind of coherent form or 'order' to emerge in this sea there has to be some way of confining where and when the singularities are most likely to occur. This is done by limiting their 'degrees of freedom' within a particular locality known as a 'probability distribution', a very common example of which is a 'bell-shaped curve' or 'normal distribution'.

Statistical form is therefore a pattern of constrained randomness, whereby the location of a chance individual occurrence cannot be pinpointed with absolute certainty, but the probability with which it will occur can be calculated according to a frequency distribution. Departure from concrete certainty or 'error' arising from separateness can hence be accounted for within a 'margin of error', *assuming that the distribution itself remains fixed*. But this assumption is not valid in an evolving system.

Whereas *statistical* modelling assumes the *intrinsic freedom* of a *multiplicity* of discrete occurrences confined within a set distribution, *deterministic* models impose a *secure* set of *initial conditions* at '*time zero*' that *completely prescribes* the *fate* of what is treated as a *singular* entity. What gives rise to unpredictability in this case is 'feedback' between 'input' and 'output', which amplifies differences in initial conditions that are too small to measure accurately into radical differences in long-term behaviour—an effect known as the 'butterfly effect'.

At the root of nonlinear theory is the paradox built into the conception of a singular numerical or geometrical figure as an independent whole, entire of itself: a 'One Alone'. The impossibility of such an independent existence was recognized poetically by John Donne, in 'For Whom the Bell Tolls':

No man is an island, entire of it self

And by William Wordsworth:

In Nature everything is distinct, yet nothing defined into absolute, independent singleness

As was demonstrated by Kurt Gödel in his mathematical formulation of the famous Cretan liar paradox, in which a Cretan informs you that all Cretans are liars, the problem is one of assuming 'completeness'. Any 'complete' or 'entire' object that thereby has nothing outside itself is inescapably self-referential and so impossible to verify or falsify 'independently' (e.g. Hofstadter 1980). All conventional mathematical proofs based on the isolation of '1' as an independent singleness are hence no more and no less than self-fulfilling prophecies.

Correspondingly, deterministic unpredictability arises quite literally from a self-referential oddity. Numbers are conceived as whole entities ('integers') and parts of whole entities ('fractions') that exist instantaneously (i.e. in *zero time*) within Euclidian dimensions of 0, 1, 2 and 3. The oddity shows up in the radically different outcomes of multiplying whole or fractional entities by themselves—what is known as 'squaring' them or raising them by the power of 2

$$0 \times 0 = 0$$
$$<1 \times <1 = \ll 1 (\text{e.g.}\, 0.5 \times 0.5 = 0.25)$$
$$1 \times 1 = 1$$
$$2 \times 2 = 4$$
$$3 \times 3 = 9$$

So, for example, $2 \times 2 > 1 \times 1 + 1 \times 1$

That is 2, as a *pair* of 1s, is more than two independent ones. The integral 'whole number', 2, is more than two independent parts of 2. Or, pictorially

$$\cdot \sim \cdot \; > \; \cdot\cdot$$

This is a very simple way of understanding the holistic concept of 'emergence', whereby a 'whole' consisting of more than one singular part is more than these parts as independent entities.

But what if the whole consists solely of fractional parts that add up to a total of 1?

A well-known, relatively simple example is provided by the 'discrete logistic difference equation', which models the pattern of growth in a self-multiplying population (e.g. Gleick 1988). This equation relates the actual number of entities (x) as a proportion of the maximum possible number (1) in a current 'population' to the number of entities in the next 'generation' (x_{next}) in terms of the net rate of reproduction (r) per head of population as follows:

$$x_{next} = rx - rx^2,$$

where x varies between zero and 1.

Here, the potential for increase in x, due to the reproductive drive, r, resulting from multiplication is countered by the negative feedback term, rx^2. When this equation is iterated (i.e. when the output x_{next} value is used repeatedly to input the next x value) from some low non-zero value, the rx^2 term increasingly inhibits the increase in x. When x is equal to $1 - 1/r$, representing the 'equilibrium population size' or 'carrying capacity' of the population, there is no further expansion.

For values of r between 1 and 3, the equilibrium population size ranges from zero to 2/3. The increase in x from low values either leads directly to attainment of the equilibrium value if $r < 2$, or, if $r > 2$, to a series of progressively smaller fluctuations (i.e. 'damped oscillations') above and below the equilibrium value. For values of $r < 1$, x diminishes to zero.

For values of $r > 3$, however, the population is driven over a threshold where it becomes unstable. Here it is unable to attain a single equilibrium state (known as a 'fixed point attractor in 'phase space'), unless arriving by some infinitesimally small chance at exactly the requisite value of $1-1/r$, and instead subdivides or 'bifurcates' into a series of alternative states. Here, as r is increased, x values come to oscillate around first two, then four, then eight ...2n values in a so-called 'period doubling' cascade. At $r = 3.57$, deterministic 'chaos' first becomes evident, as x values vary non-repetitively and at $r = 4$, all x values between 0 and 1 become possible.

Note here that the 'chaos' produced via the logistic equation is described as 'deterministic' because all the 'initial conditions' are fixed and there is a preset limit that the system cannot exceed. The system *as a whole* is effectively contained within a fixed boundary and its behaviour *can* be predicted with complete certainty so long as the initial conditions are known *precisely*. The fact that in reality the initial conditions can *never* be known precisely, and even tiny changes in initial conditions can be amplified by feedback into huge changes in behaviour, makes the behaviour unpredictable in the long run. As was recognized by Heisenberg (1927):

> [I]n the rigorous formulation of the law of causality—'If we know the present precisely, we can calculate the future'—it is not the conclusion that is faulty, but the premise.

The uncertainty in this case is not, however, due to randomness because the system is fully defined in the first place.

Notice the inconsistency here! In fact the system *cannot* be definitively closed if it is to be capable of multiplying: its boundary must be both dynamic (capable of expansion and contraction) and permeable. The supposed determinism is in the modelling assumptions using discrete numbers, not in the model itself. The unpredictability results from the inclusion of infinite space in the gaps between 0 and 1 and 1 and 2, and it is this space that makes an infinite variety of fractional values of x possible.

The exploration of a realm of infinite fractional spatial possibility within a discrete whole system is also a feature of what is known as 'fractal geometry'. This was made famous by Benoit Mandelbrot (1977), as a way to describe structures whose boundaries, unlike Euclidean lines, planes or solids, appear progressively more complex/irregular, in 'self-similar' patterns, the closer they are observed. Almost anything we examine in nature from clouds, to snowflakes, to river valleys, to ferns, to trees, to lungs has this quality. It makes them immeasurable in terms of discrete units of length, area and volume, because how much you see depends on how close you are. For example, the length of the coastline of the Isle of Wight seems much less to an astronaut orbiting the Earth than it does to a mite crawling around its many indentations. At infinitesimal scales of closeness, the perceived length is infinite.

Mandelbrot solved the problem of quantifying fractal structures by relinquishing the Euclidean premise that dimensions can have only integral values of 0, 1, 2 and 3 and allowing them also to have fractional (hence the term 'fractal') values. The fractal dimension of a structure can be calculated from the equation

$$M = kr^D,$$

where M is the material 'content' of a portion of the structure, r is the radius of the spatial field *within* which this portion of content is contained, and D is the dimension. D can readily be found from the relationship between the logarithms of M and r for different fields of view. If the structure is homogeneous, then D will have an integral value. If it is heterogeneous, D will be fractional.

Fractal patterns can be generated mathematically by iterating equations in the same way that I described earlier for the discrete logistic equation. A famous example is the 'Mandelbrot set' itself, which appeared in many guises as a colourful modern mathematical art form in the late twentieth century. This set is made by mapping the distribution of points in the 'complex plane' that do not result in infinity when iterated according to the rule, $z \rightarrow z^2 + c$, where z begins at zero and c is the complex number corresponding to the point being tested. Here, a 'complex number' is a number that consists of a combination of a 'real' and 'imaginary' component, the latter being a derivation of, 'i', the square root of -1. The complex plane is formed in the space defined by placing all 'real' numbers, from $-\infty$, through 0, to $+\infty$ along a horizontal line, and all 'imaginary' numbers, from $-\infty i$, through 0, to $+\infty i$, along a vertical line, and using these Euclidean lines as coordinates. In effect, it represents a way of increasing the 'possibility space' for numbers as discrete entities to inhabit, from one to two dimensions.

3.2   'Place-Time': The Transformational Flow Geometry of Nature

None of the above conceptual complexity and inconsistency would be necessary if it was simply appreciated in the first place that all natural form is a product of space and flux, not rigidly definable structure. And it is not difficult—although it is exceptionally unusual—to work out that there can be *no* rigidly definable structure in Nature: the very idea that there is arises from abstract rationalization.

Imagine, for example, a tennis ball. If it was not distinguishable from its surroundings, no one would be able to play tennis! In fact no one could even be around to invent or play tennis, because Nature would be featureless! Now try to imagine the tennis ball devoid of space. It can only exist as a dimensionless point, without size or shape. It has disappeared into nowhere!

Inescapably, MATTER CANNOT BE DISSOCIATED FROM SPACE. And yet the whole of rationalistic logic and mathematics, which underpins objectivistic science, is based on the misconception that matter exists as independent, self-contained particles—discrete, dimensionless 'units' or 'point-masses' that can be assembled or disassembled like 'building blocks' into smaller or larger whole objects.

If we recognize that matter cannot be dissociated from space without disappearing into nowhere, then we can recognize that for matter to exist at all, space and whatever *informative presence* distinguishes material form from its surroundings *must include each other*. And for that to be possible, this informative presence must be in *continuous motion* both within space, and around local centres of space.

Imagine yourself entering endless void darkness, but equipped with a dimensionless point of light. So long as that point of light remains motionless within the motionlessness of space it will be invisible. But move that point of light around and in due course, it will dynamically outline a transient local figure. Remember playing with sparklers on fireworks night? That is how the natural *flux* of energy brings local material form *temporarily* into being. All material forms, and all living creatures, originate in the natural inclusion of space in local flux somewhere and local flux in space everywhere. Space is a motionless presence everywhere, without limit and energy is continuous local movement somewhere in particular. Each in the other co-creates material form. Material form is born in the *mutually inclusive relationship between space and energy*, not the exclusion or conflation of one from or with the other. In the same way, all living creatures are borne in dynamic relationship with their habitat. The environment is not, as Einstein once said, 'everything that isn't me': it is the very making of 'me'.

Awareness of the mutual inclusion of receptive space and informative flux in all material bodies as *flow forms* opens up a new understanding of physical reality. We move fully from the atomistic perception and treatment of space, time, energy and matter as mutually exclusive quantities to appreciating them as mutually inclusive aspects of reality as 'place-time' (Rayner 2011). Specifically, we can recognize crucial distinctions

- between 'space' as a natural infinite presence of intangible stillness, 'place' as a distinct locality, and 'distance' as a derivative measure of the length of travel between one place and another.
- between 'energy' as a natural presence of continuous flux, and 'force' as a derivative measure of the transfer of energy from one locality to another.
- between 'matter' as a natural local embodiment of space in energy, and 'mass' as a derivative measure of amount of material.

- between 'time' as a natural inclusion of continuous flux, and 'time elapsed' as a derivative measure of the interval between one occurrence and another.

When a derivative measure is confused or conflated with the naturally continuous presence, needless complication and paradoxical reasoning occur. When 'space' and 'energy' are understood as primary natural presences that embody material form, the origin of natural patterns, processes and relationships becomes simple and non-paradoxical to comprehend.

Hence, we need to stop treating natural material bodies as if they are objects made up of discrete components driven by external forces. If we insist on thinking this way, we will never appreciate what it truly means to belong in a living world. We will perceive only the inertness of things, pushed and pulled around by ineffable externalized force.

In reality, natural flow forms and patterns obviously cannot be constructed exclusively from dimensionless points of material. They can only arise from receptive space and informative flux as mutually inclusive presences. Let us now see how.

3.2.1 Pulse and Circulation: Starting from Natural 'Spherical One', not Abstract 'Square One'

The most basic of all natural bodily forms, from which all others in the natural world arise, is no more and no less than a sphere or (if confined to a surface) circle. It is not a cube, square, straight line or dimensionless point! Herein resides the most fundamental problem of abstract science and the primarily *linear* (straight-line-based) mathematics upon which it is founded. It is impossible to derive a curve from discrete numerical units, as the 'irrational number', π, the ratio between the curvaceous circumference and linear diameter of a circle, makes evident. The *whole* of abstract mathematics begins in the wrong place, at 'square one' instead of 'circular one'!

It is not difficult to find examples of spheres and circles in the natural world: in raindrops, bubbles, eggs, blueberries, oranges, lichen colonies, mushroom caps, planets, moons, suns...to name but a few. There is something basic about this form, and we can quickly get a feel for what this is by constructing a circle on a piece of paper using a compass. We dig the needle point of one arm of the compass into the paper, making a hole, and rotate the pencil point attached to the other arm around this. The hole marks the centre of the circle as *a local point of space*, while the pencil point, as a source of informative energy, traces the continuous path of the circumference around this point. It is as though the hole, as a *receptive centre* of spatial stillness, *attracts* the movement of the pencil point into circulation around it. The movement of the pencil point cannot *coincide* with the hole without disappearing into it. Any movement of the pencil point *out of circulation*, would require an additional use of informative energy: circular form is that in which the surface-to-volume ratio is minimal. This is why circles and spheres are often said to represent the 'lowest energy state' of all natural shapes, any departure from which is energy-costly.

This is how we can understand that the starting place for all natural flow forms to emerge from is a 'circular one', not a 'square one'! Moreover, this 'circular one' *dynamically incorporates a zero-dimensional point of space* as a 'hole at its centre'. It does not exclude space as a void outside of itself, as does the '1 *or* 0' opposition of binary logic that is incorporated into digital computers and represented as juxtaposed square pixels on a video screen.

Notice also that this circular or spherical form requires the continual circulation of informative energy in order to become manifest. It cannot exist instantaneously, within zero time, because zero time precludes movement. It is an expression of *'place-time'*—the mutual inclusion of space and flux as a dynamic locality, not an inert object.

So, where, then, do those straight lines, triangles, rectangles and flat-surfaced solids, which are so easily divided up and assembled from identical sub-units of length, breadth and depth, actually occur in Nature, and how do they arise?

There are a great many *apparent* examples in Nature of the straight lines, triangles, rectangles and flat-surfaced solids beloved by abstract geometry. They occur in the superficial appearance of a huge array of frozen or crystalline solids; in multicellular bodies and organs; in honeycombs and honeycomb-like structures; in the compound eyes of insects; in *sections* cut through natural forms such as a cylindrical tree trunk; in the stripes and polygons formed on the surfaces of many kinds of plants, animals and fungi, as well as in vegetation and geological features. They also occur in the apparent flatness of a horizon, when viewed from a distance.

What all these examples have in common, however, is that they are *products*, not *ingredients* of an underlying spherical or circular geometry. Curvature comes *before* linearity in the formation of natural material bodies, not as a product of linearity, as conventional mathematics supposes.

The apparently flat surfaces of crystalline structures, from the protein coats of viruses to the hexagonal columns of basalt rocks actually arise from the underlying close-packing of arrays of tightly bound spherical molecules, analogous to a stack or gathering of cannonballs or ball bearings on a flat surface. (For example, a tight gathering of six circles around a central circle of equal diameter forms a hexagon). If these surfaces were examined sufficiently closely, the dimpling due to the underlying close-packing would become evident—so the *apparent* flatness is an illusion of observing from too great a distance. The physicist, Osborne Reynolds (1903) was so impressed by such close-packed arrays that he thought space itself is composed from them.

Honeycomb-like structures arise in a somewhat similar way through the pressing together of arrays of *flexibly* bound spheres or cylinders, which squeezes out the gaps between them. Cylindrical forms obviously arise through the elongation of spherical forms, and their parallel-sidedness is due to the maintenance of a constant diameter as they do so. If elongation is confined to or within a flat or flattened surface or structure a stripe or strap will form. If the diameter reduces or expands during elongation, a conical or parabolic form may arise. If this is confined to or within a flat or flattened surface or structure something like a leaf or flower petal will form, three or more of which will give rise to a star-shaped structure.

Whereas the apparent flatness of a crystal surface is due to observing an array of small curved surfaces from too far away to notice them, the apparent flatness of an earthly horizon is due to observing a large curved surface from too close a viewpoint. Whenever energy resides in a local body somewhere, it does so in natural circulation.

There is, however, a particular circumstance when linearity IS primary rather than secondary, and that is when energy travels as a 'pulse', directly from one locality to another locality in a 'bee-line' or 'as the crow flies'. Think of an array of dots on a sheet of paper. We can imagine a set of invisible straight lines connecting them, and indeed, using a pen or pencil we can 'join up the dots' along these lines. Between two dots we can only draw a single line. Between three or more we can enclose an area. Between four or more we can draw a network of lines. Providing the two localities are stationary with respect to each other, and there is no other distorting environmental influence, a linear trajectory or path will form when energy travels along the shortest distance between them. *Hence we can think of a linear bodily form as a product of energetic movement along an intangible axis of space, not fixed material structure.*

Correspondingly, where energy dwells locally, within a 'home range', it does so in circulatory form; where energy relocates from one place to another, it does so as a pulse (Fig. 3.1). This dynamic pattern occurs universally from quantum to galactic scales. And the combination of circulatory with linear motion opens up a vast variety of possibilities for pattern generation, not least the molecular informational form that enables particular kinds of life to reproduce themselves faithfully through successive generations, DNA.

Fig. 3.1 'Pulse and Circulation'. (Symbol of natural flow-geometry prepared by Roy Reynolds)

3.2.2 Spirals

If a cylindrical, conical or parabolic form both rotates and expands around a central point or longitudinal axis, the result will be a spiral. Spiral forms are extremely common throughout the natural world, from molecules like DNA to spirochaete bacteria, plant stems and flowers, snail shells, tornadoes, hurricanes and galaxies. They are an obvious manifestation of an underlying flow dynamic of pulse and circulation, as illustrated by this example that I created for my grandson by moving sand around *in* a beach (Fig. 3.2).

3.2.3 Ripples

While it may require some imagination to appreciate the fluid dynamic origin of spheres, polygons and spirals, it should require no imagination at all to recognize this in ripples. Who has not watched ripples spreading out as a series of concentric ridges and troughs around the entry point of a stone into a pond, or the choppy turbulence of a windblown lake or sea or a river in spate? Who could fail to appreciate that the combination of strength and weakness in liquid water, due ultimately to its mutual inclusion of informative energy and receptive space, is what enables it to respond in this way to a pulse of energy spreading through it?

Fig. 3.2 A sand-whirl.
(Photograph by Marion
Rayner)

Almost wherever we look amongst life forms, there is evidence of rhythmicity and turbulence of one kind or another. We see this in oscillatory patterns of behaviour, growth, development, physiology and bodily markings and textures. So much so, that 'rhythm' and 'life' could almost be regarded as synonymous. Given the pulse and circulation of natural energy flow, this is to be expected, but conventional mathematics has always found it difficult to deal with until, as described earlier in this chapter, it found a way inadvertently to introduce the receptivity of space and motion of energy into nonlinear dynamical systems theory.

3.2.4 Rivers and Flow-Networks

What we are used to thinking of as rivers are those branching channels that carry liquid through landscape from collection to distribution in tributaries and distributaries or estuaries, via a main channel. Flow within this channel rushes or meanders along its course depending on gradient and local resistance to movement. The lowering of resistance to current by erosion accelerates the flow along deepening channels, while sedimentation in slack current increases local resistance in shallowing channels.

This same, basic flow pattern also occurs, however, in a vast variety of biological structures. These include nerve cells, blood vessels, fungal hyphae, leaf veins and trees, as well as transportation routes between sites of collection and distribution in foraging and migrating groups of organisms and even in 'stream of consciousness' thought processes. All arise through the assimilation of sources of energy through the surface of an elongating boundary at a rate faster than can be accommodated by a single growth point.

Tree-like or 'dendritic' branching patterns result when branches diverge from one another to yield a radiating system. If leading trunks or branches exceed the growth of lateral branches, a Christmas-tree-like, conical pattern arises. If lateral branching exceeds axial outgrowth, a more rounded pattern results.

When branches *converge*, however, they may actually make contact with one another and fuse. The result of convergent branching and fusion in a river system is known as anastomosis, and this can occur also in a wide variety of biological structures. It converts the initially dendritic system into a *flow network*, which enables energy to be *cycled* and recycled as well as to flow back and forth. Moreover, it replaces a set of pipelines that are connected *in series*, with a set that is connected *in parallel*, which has a much lower overall resistance to current flow.

Flow networks are hence especially important in the formation of distributive and circulatory communication systems. An instructive example occurs in the mycelia of higher fungi. These are the behind-the-scenes 'production teams' responsible for the amazingly rapid emergence of mushrooms and toadstools. The following photograph shows the mycelium of the Magpie Ink-cap, *Coprinopsis picacea*, grown in a laboratory culture through an array of alternating high and low nutrient-containing chambers (Fig. 3.3).

Fig. 3.3 'Sustainable development' in abundance and scarcity, illustrated by mycelial growth of the magpie fungus, *Coprinopsis picacea*, in a matrix of 25 2 × 2 cm plastic chambers filled alternately with high and low nutrient media. Holes have been cut in the partitions just above the level of the medium. The fungus has been inoculated into the central high nutrient chamber, whence it has produced alternating prolific and condensed patterns of development. Growth linking between chambers has been reinforced into persistent 'cables', whereas mycelium unable to extend further has been prone to degenerate. (Photograph reproduced by courtesy of Louise Owen and Erica Bower)

Here, it is worth noticing that flow networks are very different both in the way that they form (as a system of expanding *tubes* or *channels*, that branch and anastomose *as they grow*), and in their function, from another kind of network, whose function is to *trap* sources of energy at a particular locality within a *meshwork* of intersecting threads. Spider's *webs* are well-known examples. I will discuss this further in Chap. 5.

3.2.5 'Crazy Paving'

I have already mentioned how equal and even outgrowth from neighbouring localities can give rise to polygonal arrays. Where outgrowth is uneven and unequal, however, much more irregular patterns can form, which resemble adjacent

Fig. 3.4 Mosaic of individual colonies of the lichen, *Opegrapha gyrocarpa*, growing on a gravestone. (Photograph by Marion Rayner)

countries on a map, or crazy paving. An easily observable example occurs in 'lichen mosaics', where adjacent lichen colonies of the same and/or different species come into contact (Fig. 3.4).

A cross-section through a piece of decaying wood often has this appearance due to territorial 'zone lines' formed between adjacent fungal mycelia. On the other hand, the meeting places known as 'sutures' between the plates that form a skull, and those known as 'tessellations' between adjacent cells that make up the epidermis of a leaf are joins rather than demarcations.

References

Einstein, A. (1954). *Relativity* (p. 138). Methuen & Co: University Paper Back, London.

Gleick, J. (1988). *Chaos*. London: Heinemann.

Heisenberg, W. (1927). Über den anschaulichen Inhalt der qauntentheorestischen Kinematik und Mechanik. *Zeitschrift fur Physik, 43,* 172–198.

Hofstadter, D. R. (1980). *Gödel, Escher, Bach: An Eternal Golden Braid*. England: Harmondsworth.

Mandelbrot, B. (1977). *The fractal geometry of nature*. New York: Freeman.

Poincaré, H. (1905). *Science and hypothesis*. Dover Publications. Walter Scott Publishing Company Ltd.

Rayner A.D.M. (2011). *NaturesScope: Unlocking our natural empathy and creativity - an inspiring new way of relating to our natural origins and one another through natural inclusion*. O Books.

Reynolds, O. (1903). *On an inversion of ideas as to the structure of the universe (The Rede Lecture, June 10, 1902)*. Cambridge: Cambridge University Press.

Smith, B. (Ed.). (1988). *Foundations of Gestalt Theory*. Munich and Vienna: Philosophia Verlag.

Chapter 4
Natural Inclusion and the Evolution of Self-identity

Abstract There is abundant direct and indirect evidence that life and the cosmos have not remained unchanged since an act of instantaneous creation. Life and the cosmos evolve through an ongoing cumulative process of energetic transformation that is evident even within the life spans of individual organisms. Efforts to understand this process have, however, been impeded by the same abstract dissociation of matter from space that sustains belief in instantaneous creation. The resultant view of evolution as an eliminative 'survival of the fittest', brought about by a selective mechanism is both paradoxical and pathological. Selection from amongst a set of competing alternatives in a fixed arena cannot initiate change and it removes diversity. A more realistic understanding arises from appreciating natural flow-geometry. Energy flows naturally in response to the inductive influence of receptive space. Life and individual identity hence evolve through the natural inclusion of space in flux (and flux in space), not the independent existence of each from the other. New varieties arise through the opening up of new possibilities for energetic expression in continuously changing circumstances, not the illusory separation of chooser from choices that arises from objective perception.

4.1 Discrete and Continuous Perceptions of Natural Origins and Destiny

Whether we realize it or not, the way we human beings live and experience our lives is profoundly shaped by the way we think about questions concerning our natural origin and destiny. Has life always been as we experience it currently, or has it changed? If it has changed, will it continue to do so, and, if so, how? Where has life come from, and where is it going to and how much control do we really have over the direction it takes?

I suspect many, if not most of us prefer not to think about these questions. We simply want to get on with our lives as contentedly and painlessly as we can in the circumstances that we find ourselves in. Other creatures appear to do so, so why should not we?

There is, however, a great and perhaps uniquely human problem for us in avoiding personal reflection on these questions. If we do not, others may seize the opportunity to do our thinking for us. We may then become dependent on these others as 'Authorities', who gain power over us that can all-too-easily be abused (see Chap. 6).

If we start with the assumption that matter exists independently from space and time, or that it all began in a moment of creation, we have no need to consider the natural origin and destiny of our universe. We have been led to believe either in eternal material existence or in a discrete act of instantaneous creation of matter from no matter—'something from nothing'. This is why it appears as if the only way in which change can be effected is through the assembly, disassembly and re-assembly of discrete atomic particles of matter into different combinations by some organizing force or administration. But how such an administration (i.e. the Law of Nature) actually works, where it is situated, and where and how it itself originated and is destined are all inscrutable. The inescapable gap in our knowledge is often filled by creation and destination stories that we are expected to believe in or not, depending on which authority we follow. Such stories have been a persistent feature of human thought throughout recorded history, but they all lack supporting evidence or sound reasoning based on actual experience.

There is, on the other hand, abundant evidence and good reason to accept that Nature has not remained unchanged, but has instead transformed due to an ongoing evolutionary process. I will not rehearse all this evidence because much of it is well known and readily credible. It comes from scientific studies of fossils, geographic distribution patterns, comparative anatomy, embryology, genetics and astronomy. Moreover, we experience personal development within our individual lifetimes as our bodies mature, our circumstances change and our thought processes are transformed by new learning.

What I do question is why our understanding of this evolutionary process continues to be impeded by adherence to that same mental dissociation of matter from space that sustains belief in instantaneous creation. The 'Battle of Ideas' between 'Darwinism' and 'Creationism'/'Intelligent Design' continues to fuel adversarial debate to this day. But in reality Darwinism merely substitutes one kind of external, judgemental administrator for another, referred to as 'Natural Selection, or the Preservation of Favoured Races in the Struggle for Life' (Darwin 1859).

But I am not only going to point out problems with the concept of 'natural selection'. I will also offer a different way of understanding of how natural patterns of life and cosmological organization originate and transform *continuously*, through 'natural inclusion'.

4.2 Paradoxical Evolution: Abstract Theories of the Origin and Transformation of Life as an Independent Existence

There is a radical difference between reasoning that something must *come from* somewhere, and believing that something must *start* somewhere. Correspondingly, theories of biological and cosmological evolution from a discrete material origin arouse unanswerable questions. Where did the material come from in the first place? What was present before the start? How do natural variability and diversity arise?

Modern Darwinian and neo-Darwinian theories either do not concern themselves with these questions, believing them to be beyond the remit of rational scientific enquiry, or offer explanations based on their own presuppositions of independent material existence. Hence a 'starting point', or set of starting points is assumed, and all else is construed from there. It is rather like being presented with a set of Lego blocks as 'units of information', and assembling evolutionary variety from these, based on a system of 'Design Rules' or 'Laws of Nature' that prescriptively determine what is and is not permitted. In modern evolutionary biology, the discrete 'units of information' are perceived to be 'genes'—sequences of DNA that encode different kinds of protein structure. In modern cosmology, these units are perceived to be sub-atomic particles arising from a primordial 'big bang' out of nowhere.

A pertinent illustration of this kind of thinking is provided by a recent experiment designed to identify the 'smallest genome' necessary to enable bacterial cells to live an autonomous (i.e. independent) existence (Hutchison 2016). The results of this painstaking work were hailed as a 'breakthrough' in recognizing what needs to be known in order to understand the origin of life, because a third of the 'essential' genes had no known function. If the function of those genes could be identified, the researchers reasoned, we would be well on the way to understanding what makes the independent existence of life possible. But what did the study really show? Actually, it showed that there can be no such thing as an independent life because life itself originates in the mutually inclusive *relationship* between an organism and its natural neighbourhood (Marman and Rayner 2016). How so?

The actual experiment started with bacteria that had the smallest genomes known. The experimenters then deactivated the genes present in these genomes, one at a time. If the bacteria lived and kept reproducing, those genes were deemed unnecessary and removed. After 20 years the genome was reduced to half its original size. None of the remaining genes can be eliminated.

But there's more to the story. Many of the 'unnecessary' genes could only be deleted after supplying the petri dish with key nutrients and eliminating potential dangers. As a result, the new cells can no longer survive in the wild. Is it fair to say that these are independently living cells? Do not they need the experimenters to feed them and remove their wastes?

Moreover, while the genomes of these cells may be small compared to other single-celled organisms, they are still 200 times larger than the genomes of simple viruses. Viruses are not, however, considered to be independent life forms because they need a host cell in which to live, and they need the genome of the host cell to reproduce. What is so different about coming alive within a host cell, and coming alive within a receptive environment?

All organisms depend on their environment for energy, carbon, water and mineral nutrients to grow and reproduce. No plant, animal, or microbe can thrive without this supply. Cutting them off leaves them as inactive as a car without fuel.

All biologists know this. But if we consider its implications deeply, it shows that treating organisms *as if* they are self-contained, genetically determined entities, isolated from their neighbourhood, is a fundamental mistake because life is not a property of individuals alone.

When DNA is removed from cells, the cells can live on for a while, but the DNA stops participating in life. It becomes an inactive chemical compound. Clearly DNA does not possess life by itself. But if life is a relationship between a life form and the world it is nourished by, then yes, DNA comes alive when inside a cell, just as cells come alive when in a nurturing environmental context, and seeds 'break dormancy' when supplied with the moisture they require to germinate and grow into mature plants. If genes become alive when they are involved in the life of a living cell, the same can be said for all the other constituents of cellular life, not least the proteins whose two-way relationship with DNA is so central. Every day, biologists see the liveliness of enzymes, as they sustain the life of the cells they inhabit. The origin of life reaches all the way down to molecules and atoms, as long as they reside in the right environmental context.

All that is overlooked, however, if life is assumed to originate and evolve from genes as 'independent units of selection' that alone determine the character of the cells, organisms and even the habitats that they inhabit. This neo-Darwinian assumption featured prominently in the emergence of eugenic principles as a means of genetic betterment of the human race during the twentieth century, as well as in the rise of 'Sociobiology' (Wilson 1998) during the 1970s, and its most celebrated and notorious idea or 'meme', the 'Selfish Gene', advocated by Richard Dawkins (1989). It also contributed significantly to the emergence of monetarist policy during the 1980s and global advance of materialistic consumerism that continues unabated (Gabriel 2002). Whilst some people, even its proponents, have lamented the unpleasant implications of this belief as a source of profound intolerance, conflict, distress, exploitation, oppression and waste, its underlying logic has seemed impossible to refute. Indeed, Richard Dawkins (1989) has urged 'If you wish, as I do, to build a society in which individuals co-operate generously and unselfishly towards a common good, you can expect little help from biological nature. Let us teach generosity and altruism, because we are born selfish.' In effect, human conscience is here called upon to overturn the tyranny of genetic self-determination so that we can behave more nicely towards one another as members of a group!

The much vaunted argument that natural selection is the *mechanism* of evolution is not, however, consistent even with the abstract logic of Newtonian mechanics (Marman 2016)! How can a post hoc choice of what is best *drive* change in the same way that a hammer drives a nail into wood? It cannot: the natural generation of variation has to come first (Rayner 1997). And how could the selective removal of variation result in evolutionary diversification? A crucial aspect of evolution is clearly missing from the argument.

Moreover, to argue that it is natural to be selfish based on a model of evolution by individual selection presupposes that it is possible to define 'self' as an autonomous unit or 'whole' that either 'is' or 'is not' (Rayner 2011a). Equally, the altruistic surrender of local individual identity and agency to collective identity (as in 'group selection') simply transfers the definition of autonomy to a larger 'whole' or 'unity' (e.g. from 'individual' to 'family', 'race' or 'species'). Both kinds of perception are deeply questionable. They can only hold true if it is possible to *cut* space by inserting a complete boundary limit or definitive hard line between one individual or group identity and another. There is no evidence for this and, as was recognized by Baldwin (Valsiner 2009), it cannot make sense in any evolutionary (irreversibly changeable) system (see also Rayner 2004). Evolution itself is contradicted by the supposition of self- or group exclusiveness upon which 'selection' theory is founded. A completely closed system has no capacity for change or relationship with any other—a point recognized by Bertalanffy (1968), but problematic to address satisfactorily within the definitive framing of general systems theory. Self-or group-preservation implies the fixture of life in suspended animation, not the evolution of life as a creative flow. True evolutionary innovation involves sustaining life as a flow of energy, not preserving life forever in a jam-jar!

4.3 Evolving Out of the Box: Life in Dynamic Relationship

The abstraction of material content from spatial context implicit in Darwinian and neo-Darwinian theory has the effect of reducing material to a set of independent particles and space into a fixed container or 'niche' that the particles compete amongst themselves to fit into. This competition culminates in the triumph of the 'winner' or 'most favoured'. Like a 'first past the post' election or a TV talent contest, it is a process that *removes* the diversity that it relies upon to get started. It is a route from diversity to monopoly, not monopoly to diversity. The fact that 'competition' is widely equated with 'diversity' is correspondingly one of the great absurdities of modern capitalism as an expression of 'social Darwinism'. Competitive opposition with the intent to defeat 'other' does not drive evolution, it stalls it.

Darwin based his idea of natural selection on the work of plant and animal breeders aimed at producing varieties of life that were more suitable for particular purposes than those in the wild. He then simply projected his perception of Man as an independent selector, keeping what he most favoured for breeding purposes and discarding the rest, onto Nature as an independent selector doing likewise. That projection was itself clearly based on separating Man from Nature, and hence Nature as the Chooser separate from its Choices, as mutually exclusive entities locked in power struggles. The resultant focus on selection as instrumental led him and his followers to overlook the real lesson that could have been learned through an appreciation of natural flow-geometry. The real lesson is that the evolution of varied patterns of life is a continuous dynamic process that depends on *making openings* for creative expression, not closing them down (Rayner 1997). Through breeding, human *participation* in the natural evolutionary process opens up the space of the habitat, making it possible for new variants to come into being. This is a *process* of energy being *induced* to flow into receptive void in the environment, not a *mechanism* driven by external force.

The biologists who engaged in the 'smallest genome experiment' described earlier actually created the local spatial opportunity for a new variety of life to emerge by removing resistive influences present in the wild, the same as plant and animal breeders. Their search for novelty *opens up* new channels for evolutionary development through *receptive acceptance* of what emerges. Their appearance of breeders *choosing* what they want from a set of competing alternatives, and the idea that this causes evolution, is an illusion. Compare this to the idea of organisms *yearning to find what they want*—their desire to thrive—as the true inspiration.

Someone shopping for clothing in a department store is *looking for what is fitting*, not forcefully discarding what is not. Once they find something suitable, it flows from shelf to empty shopping bag where it is gratefully received. It is this subtle but radical difference between selective do-or-die competition and receptive acceptance and responsive flow that distinguishes open-ended evolution by natural inclusion from the closed-system sterility of Darwinian selection. The receptive and empty feeling of 'what-is-wanting' brings about evolutionary innovation beyond the comprehension of objective reason.

Recognizing that material content *cannot* be isolated from space as an independent occurrence, and that material bodies themselves arise from the mutual inclusive relationship between receptive space and informative flux, hence offers an understanding of evolution that is far more consistent with actual experience:

> Life and the cosmos evolve through the receptive natural inclusion of what's possible in changing circumstances, not the competitive elimination of what's impossible in a fixed arena.

Let us now consider this further. We can begin by recognizing that to be entirely self-contained is to be an inert, hermetically closed structure with no capacity for take up or loss of energy between inner world and outer world. This 'stay-inside' structure keeps its circulating energy to itself in permanent residence: it never emerges out from its permanently closed doors. The nearest any life forms actually get to this condition is when they form survival capsules such as spores, seeds, pupae and cysts that carry them through periods of scarcity. This is what real biological 'survival' or 'preservation' entails. In such a dormant condition, they are incapable of any active growth or relationship with others. But no sooner is any activity resumed that can support growth, so too is any life form's capacity to lose as well as take up energy through its necessarily permeable bodily boundaries and those of others in its vicinity.

Fig. 4.1 'Fountains of the Forest' (oil painting on board, by Alan Rayner 1998)

The availability of sources of energy is hence the principal influence that governs the growth, organization and function of all forms of organic life as variably open systems. Any activity or pattern of development in which energy loss through permeable boundaries persistently exceeds energy acquisition will result in unsustainable deficit. On the other hand, anything that permanently prevents energy loss also prevents energy gain (Rayner 2011a).

For any living system to sustain itself, its primary need is therefore to be able to *attune* its activities and development to *correspond with* energy availability and hence with the local conditions of its habitat. This availability varies in amount and rate of supply due to seasonal and climatic fluctuations, and where and in what form it is located. It also changes due to the growth, death and decomposition of the systems themselves, which respectively deplete and replenish supplies as they come under one another's simultaneous mutual influence. For example, as I once fluidly described and illustrated (Rayner 1998; Fig. 4.1), within a forest, 'a tree is a solar-powered fountain, its sprays supplied through wood-lined conduits and sealed in by bark until their final outburst in leaves … Within and upon its branching, enfolding, water-containing surfaces, and reaching out from there into air and soil are branching, enfolding, water-containing surfaces of finer scale, the mycelial networks of fungi … which provide a communications interface for energy transfer from neighbour to neighbour, from living to dead, and from dead to living'.

Real life does not inhabit an even playing field of energy, space and time. Instead it continually both changes and responds to changes in the contextual circumstances of its natural neighbourhood in an improvisational process that gives rise to evolutionary and ecological complexity and succession (Rayner 1997, 2004). This process of natural inclusion is *co-creative*, transforms all through all (Rayner 2011b), and opens the way to extraordinary diversity and complexity of interdependent forms and patterns of life as they co-evolve over myriad nested temporal and spatial scales. The breath-taking variety that we can find in a crumb of soil, a patch of chalk grassland, a coral reef and a tropical forest comes into being under the guidance of no more and no less than the responses and contributions of its membership to natural energy flow in a natural 'sustainability of the fitting' (Rayner 2010; cf. Elstrup 2009).

This is why, as depicted in Fig. 4.2, the boundaries of real organisms, populations and communities vary in permeability, deformability and contiguity

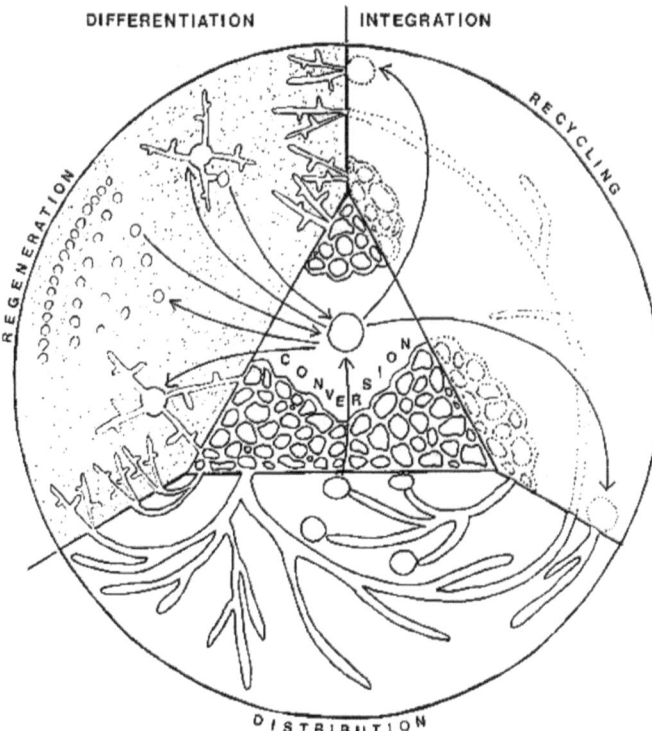

Fig. 4.2 The interplay between boundary-differentiating and boundary-integrating processes in energy-rich (stippled) and energy-restricted circumstances. This interplay enables energy to be assimilated (allowing regeneration and proliferation of boundaries), conserved (by conversion of boundaries into relatively impermeable form), explored for (through internal distribution of energy) and recycled (via redistribution/reconfiguration of boundaries) in spatial capsules, channels, branches and networks of life forms in dynamic attunement with their natural neighbourhood. Thin lines indicate relatively more permeable boundaries, thick lines relatively impermeable boundaries and dotted lines degenerating boundaries. (From Rayner 1997)

(connectivity) (Rayner 1997; Elstrup 2010). They *reconfigure* in dynamic relationship with the availability of energy predominantly assimilated from sunlight into organic compounds via the process of photosynthesis, and rendered into chemical form (adenosine triphosphate) via the oxidative-reductive reactions of respiration as a form of combustion. Moreover, these reconfigurations themselves entail alterations in boundary chemistry induced by the availability and production of oxidizing and reducing power (Rayner 1997; see Chap. 5).

The ecological and evolutionary sustainability of natural life forms, from the cells and tissues in a human body to the trees in a forest depend upon close mutual *attunement with* (as distinct from unilateral *adaptation to*) the diversity, complementary nature and changeability of all within their neighbourhood, *to which they themselves contribute*. When energy supplies become scarce, sustainable living systems pool and redistribute internal resources within integrated structures and survival capsules—they do not compete to proliferate faster on the dwindling supplies than their neighbours. When supplies are abundant, they proliferate and differentiate. Moreover, as is beautifully illustrated by the exploratory patterns of some kinds of fungi, this ability to attune their capacity to differentiate and integrate activity in dynamic relationship with energy availability allows life forms to locate and sustain supplies in heterogeneous habitats with extraordinary efficiency. As illustrated in Fig. 4.3, they do this through a combination of all- round exploration and directional focus (i.e. 'circulation' and 'pulse', as described in Chap. 3).

Figure 4.3 shows how the mycelium of the wood-decaying fungus, *Hypholoma fasciculare*, finds an 'oasis in a desert', by fluid-dynamically spreading and narrowing its energetic focus. The fungus has been inoculated into a tray full of soil on a block of wood ('starter' food source), with an uncolonized wood block ('bait' food source) placed some distance away from it. Distinct stages are shown in the radial spreading of the fungal colony from the inoculated wood block, followed by the redistribution and directional focusing of its energy following upon contact with the bait. As indicated in Fig. 4.2, similar fluid dynamic patterns of gathering in, conservation of, exploration for and redistribution of energy supplies within variably connective channels and capsules of receptive space are found throughout the living world, from subcellular to ecosystem scales of organization.

Sustainability, not supremacy, is the path of evolutionary and ecological continuity. Natural energy flow is variably fluid, circulatory and redistributive along pressure gradients from higher concentration (relative 'abundance') to lower concentration (relative 'scarcity'), as in atmospheric and ocean currents. The primary need for all life forms is not to seek competitive advantage through the unilateral accumulation of energy 'wealth' at the expense of their neighbourhood, but to sustain themselves and their offspring as variable channels for natural energy flow. They are more like members of a relay team, continually receiving, temporarily retaining and eventually passing along what sustains life, than a set of autonomous individuals striving to be first past the post. To succeed in this they have to be open

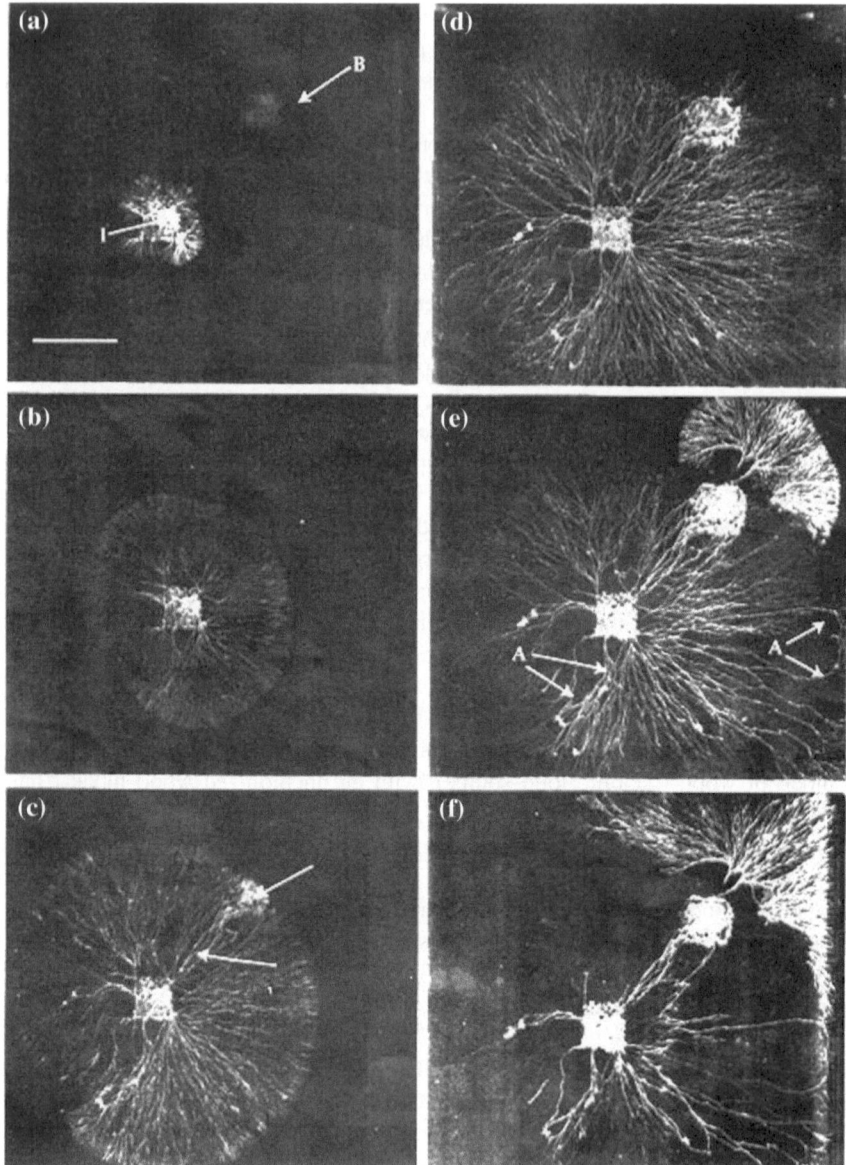

Fig. 4.3 'Fungal Foraging' (from Dowson et al. 1986; see also Rayner 1997)

to the energetic influence of their neighbourhood at the same time as sustaining the distinctiveness, but not separateness, of their inner worlds from their outer worlds through their dynamic boundaries.

Any ecological or evolutionary model that treats an individual or group as a discrete, autonomous object or subject with the set objective of promulgating and preserving its self as sole survivor of a war of attrition is therefore inapplicable to a changeable world of natural energy flow. Yet just such treatment underpins the Darwinian concept of 'natural selection' as 'the survival of the fittest' or 'preservation of favoured races in the struggle for life' (Darwin 1859).

Why, then, does this concept persist?

4.4 From Self-definition to Self-opening

Perhaps the continuing popularity of Darwinian thinking is due to the fact that it both arises from and reinforces a deeply rooted perception of self-identity that is especially prevalent in modern North America and Western Europe. According to Walker (2003):

'Cross-cultural views of the self define individuality in terms of boundaries, locus of control and inclusiveness versus exclusiveness, or that which is intrinsic versus that which is extrinsic to the self (Heelas and Lock 1981; Sampson 1988). Cultures that emphasize firm boundaries and high personal control tend to view the self as exclusionary or "self-contained." Fluid boundary, strong field control cultures, view the self as "ensembled", meaning that the self is inclusive of other individuals. While "self-contained" individualism is indigenous to the United States and to the European countries from which its dominant ethnic groups draw their roots, "ensembled" individualism is far more prevalent as a percentage of all known cultures (Sampson 2000). Ensembled individualism is also indigenous to Aboriginal, Native American, Senoi and other cultures that are widely known to use dreams for social purposes.'

Clearly natural inclusion corresponds with the ensembled individualism and fluid boundary perception of indigenous cultures, which, I might add are characterized by an outdoor, nomadic life style that combines 'pulses' of migration with settled 'circulation'. Perhaps my own personal experience of 'culture shock', described in Chap. 1, arose from the contrast between my immersive outdoor childhood experience in Kenya and my rationalistic English urban education where all serious study was boxed indoors and learning experience 'outdoors' reserved for 'holidays' and 'playtime'.

Definitive perceptions of self-identity exclude self from neighbourhood, associated with a strongly self-assertive, narcissistic psychology. As C.S. Lewis (1942) put it in 'The Screwtape Letters' from a senior to an apprentice devil:

The **whole** philosophy of Hell rests on a recognition of the axiom **that one thing is not another thing**, and, specifically, **that one self is not another self**. My good is my good and your good is yours. What one gains another loses. Even an inanimate object is what it is by excluding all other objects from the space it occupies; as it expands, it does so by pushing all other objects aside or by absorbing them. A self does the same... 'To be' means 'to be in competition'. Now the Enemy's philosophy is nothing more or less than one continued attempt to evade this very obvious truth.... He is not content, even Himself, to be a sheer arithmetical unity; He claims to be three as well as one, in order that this nonsense about Love may find a foothold in his own nature... The whole thing, in fact, turns out to be simply one more device for dragging in Love.

Notions of adversarial 'competition' and coercive 'co-operation', which respectively underlie individualistic 'capitalism' and collectivistic 'socialism', are predicated upon this axiomatic logic (Rayner 2011a). It is presupposed that individual or group identities can be defined independently from their spatial context and correspondingly that their 'future' can be fully defined by present or 'initial conditions' (cf. Chap. 3). As recognized by Bateson (1972), this can give rise to the familiar idea that undesirable present 'means' can be justified by desirable future 'ends'. Human beings may be cognitively and culturally predisposed to make this presupposition through a combination of our inter-related capacities for categorization, sociality, abstract thought, tool and language use and awareness of mortality (Rayner and Jarvilehto 2008; cf. Elstrup 2009, 2010). On the other hand, the imagination that comes alongside these capacities offers us the possibility of escaping restrictions of objectivity through the more comprehensive worldview of natural inclusionality (Rayner 2011a, b).

As terrestrial, omnivorous, bipedal primates unable to digest cellulose but equipped with binocular vision and opposable thumbs that enable us to catch and grasp, we are predisposed to view the geometry of our natural neighbourhood in an overly definitive way. We then see the world in terms of what it can do for us as detached observers, not how we are inextricably involved in it as natural inhabitants. We perceive 'boundaries' as the limits of definable 'objects' and 'space' as 'nothing'—a gap or absence outside and between these objects (Rayner 2004).

As I have already mentioned, this perception of space and boundaries as definitively discontinuous renders the comprehension of continuity and change problematic (see also Smith 1997). If two adjacent locations in space and/or time are distinguished by a boundary, which one does the boundary belong to? If it belongs to both of them, how can the mutual exclusivity of definitive logic be satisfied, and where do both cease to be both and become either one or the other? If it belongs to neither, then where does one location end and the other begin and what really comes between them? In the case of a curved boundary, does it belong to whatever lies within it or to whatever lies without it? If two distinct locations are both contained within a larger location, are they mutually exclusive or co-existent? Upon such dilemmas rests the whole gamut of alternative propositional (either/or) and dialectical/transcendental logics (both/and in mutual opposition) that have been in conflict for millennia and continue to be so (e.g. see Valsiner 2009). So too do the 'holons'—as 'Janus-faced' entities combining individual and collective aspects, and 'holarchies'—as nested arrays of holons, of Koestler (1976) in his 'Open Hierarchical Systems Theory' (cf. Wilber 1996).

That it is nonetheless possible to avoid this perception is evident from the indigenous cultures that sustain a much stronger sense of inclusion in Nature, aided by the preservation of oral, aural and nomadic traditions (e.g. Cairns and Harney 2004; Taylor 2005). For example, notice the similarity between the following quotes from Bill Yidumduma Harney (BYH), a fully-initiated Elder of the Wardaman people of Northern Territory, Australia (see Cairns and Harney 2004) and a 'natural inclusional poem', 'The Hole in the Mole', by myself (AR).

BYH: 'You might recognize some of the land, changing all the time. Then, like imagination to us, with spiritual link-up from the stars, and all the other stuff from the top to the bottom, they sort of guide you all the way. They start like be still in the valley, you have got it in your mind, links the air to you, up to the stars, guide you direct to it straight across country... all these stars pulling everything together, moving around, all come together'.

AR: 'The Hole in the Mole'

'I AM the hole; That lives in a mole; That induces the mole; To dig the hole; That moves the mole; Through the earth; That forms a hill; That becomes a mountain; That reaches to sky; That pools in stars; And brings the rain; That the mountain collects; Into streams and rivers; That moisten the earth; That grows the grass; That freshens the air; That condenses to rain; That carries the water; That brings the mole; To Life.'

All that is needed to open up our self-identity from the unnatural confinement imposed by definitive rationality is the simple understanding that space cannot be cut, occupied, displaced, confined or excluded. Space is a continuous, intangible presence throughout and beyond the boundaries of natural figures (Rayner 2011a). By the same token, these boundaries are energetic interfacings between inner and outer realms, not fixed limits. This simple move from regarding space and boundaries as sources of discontinuity and discrete definition to sources of continuity and dynamic distinction is the ecological and evolutionary point of departure of 'natural inclusionality', as a philosophy of natural inclusion, from objective rationality.

The following simple exercise might help illustrate the difference between the hard line, space-cutting view of definitive logic and the fluid-lining understanding of natural inclusion (Rayner 2011a). Draw an outline of two figures using a dotted line on a plain sheet of paper, as in Fig. 4.4. The 'paper' infinitely stretched would represent what in the transfigural geometry developed by Lere Shakunle is called 'Omni-space' (Shakunle and Rayner 2008, 2009). The space within each figure represents 'Intra-Space', the space between figures 'Inter-space', the space beyond the figures 'Extra-space' (not labelled in Fig. 4.4) and the space transcending the figures' permeable and dynamic boundaries 'Trans-Space'. You can see how the continuous non-local space everywhere ('Omni-space') is configured into distinctive, but not discrete dynamic localities. In the way that you have drawn them, the figures are not contiguous (connected), and so they can only communicate through

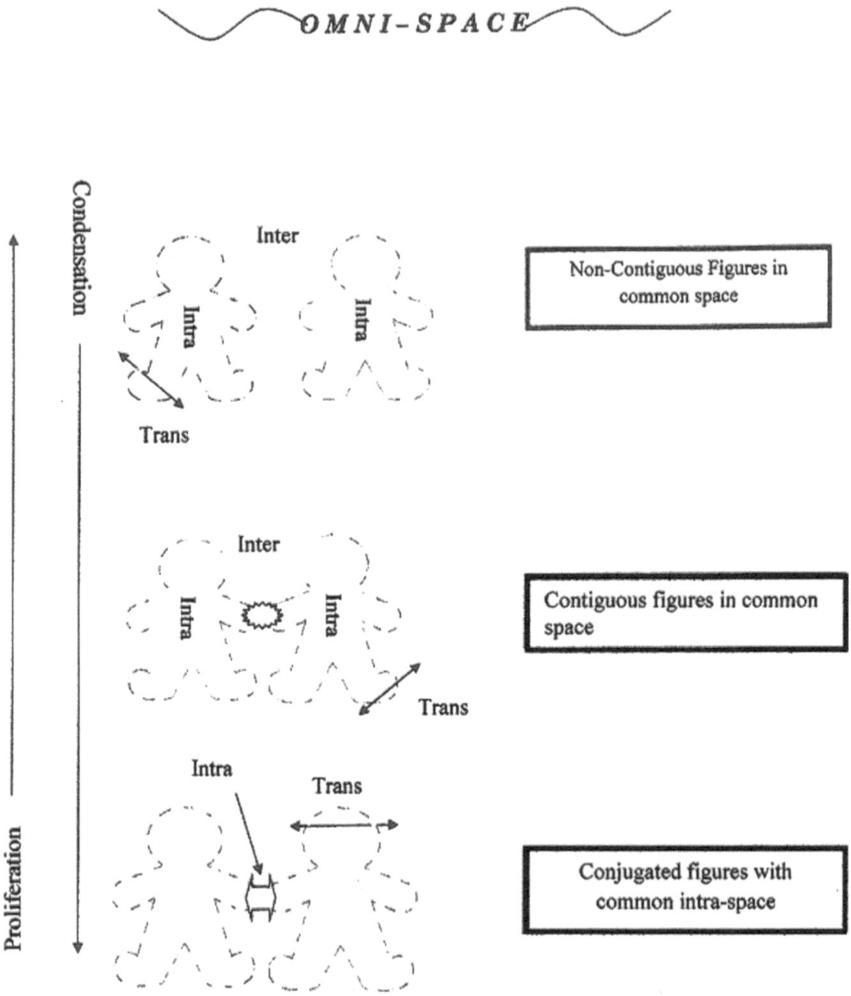

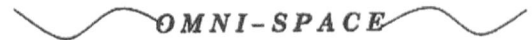

OMNI-SPACE

Fig. 4.4 Distinct but not discrete figures of space in space (from Rayner 2011a)

the 'inter-space' and 'trans-space' between and permeating their boundaries as energetic interfacings and restraining influences. Nonetheless, they inhabit the same limitless pool of omni-space everywhere. If you were now to draw the figures closer together, so that their boundaries first connect and then coalesce at one or more points, their intra-space now becomes continuous. On the other hand, if you were to take a pair of scissors and cut around the dotted lines, the figures will drop out of their spatial context as discontinuous individual entities. This 'dropping out' of context is what discontinuous models of reality effectively do—they treat boundaries as cut-out zones between discrete inner realms and outer realms, instead of dynamic relational interfacings through which these realms remain in continuous communication through trans-space.

Figure 4.4 hence illustrates the dynamic relationships between figural flow-forms as energetic configurations of space. It also serves to distinguish the natural–inclusional relationship between distinct but not discrete flow-forms both from reductive schemas that cut off inner from outer spatial realms and from connective and holistic schemas where individual dynamic locality is eschewed from a seamless, purely figural whole or 'unity'. Since the cartoons can only represent an instantaneous 'slice' through the figures, the dotted lines should not be taken to represent 'sieves' but more the kind of seething 'fluid mosaic' that constitutes real biological membranes (see Chap. 5). A very simple example of what is represented in the cartoon can also be seen between surface-tense droplets of water condensing on a surface. As they expand and come into proximity their tensely curved inner–outer interfacings first touch and then coalesce in a visible rush as each flows reciprocally into the other and the tension of their boundaries is released. A living illustration of the process of figural boundaries coming into proximity, contiguity and conjugation occurs during the process of hyphal fusion that is found in many fungi (e.g. Ainsworth and Rayner 1986) and is shown in Fig. 4.5.

Here we can see clear, fundamental differences between rationalistic and natural inclusional perceptions of connectivity and continuity:

1. In rationalistic thought, continuity is *equated with* 'connectedness' because space is regarded as a source of discontinuity/gap between and around 'things' as discrete objects. Hence the only way of deriving continuity in this 'whole way of thinking', is either by totally excluding space and boundaries from form as a continuous line or network of width-less threads, or by totally conflating space with form in a seamless whole. Such exclusion or conflation is neither consistent with evidence/experience nor does it make consistent sense.

2. In natural inclusional thought, space is a continuous omnipresence that cannot be cut, occupied, confined or excluded, and form is *dynamically continuous* through its energetic inclusion of space throughout figure and figure in space. Distinction and difference are hence accommodated in a natural fluid continuum, *without contradiction*. Local identity is recognized as a dynamic inclusion of non-local space in which all forms are pooled together (but not merged into complete unity) in natural communion as flow-forms.

3. Correspondingly, the treatment of continuity by objective rationality as the same as connectedness is an idealized abstraction that is physically impossible. The very idea of complete 'whole units' existing anywhere, at any scale in Nature as an energetically open, fluid system does not make sense. The *fluidly variable connectivity* of natural inclusionality arises from the coming together, fusion and dissociation of energetic paths, corridors or channels of included space in

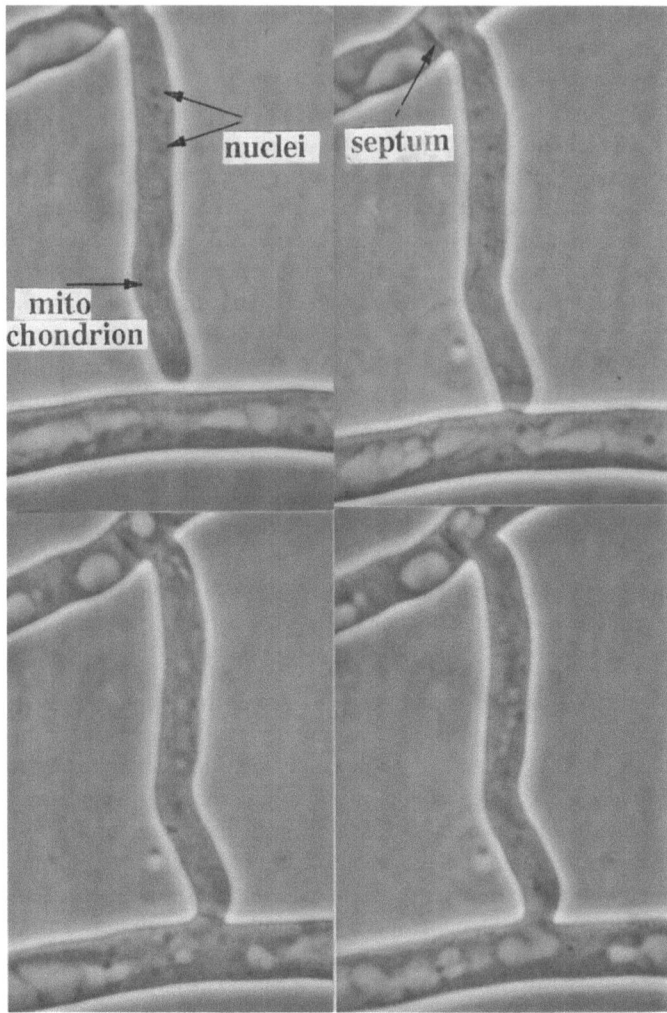

Fig. 4.5 Stages (from *top left clockwise*) in fusion between the protoplasm-filled cellular tubes (hyphae) within the mycelium of the basidiomycete fungus, *Phanerochaete velutina*. The tubes are internally partitioned into distinct compartments by septa, which have a door-like pore in their middle. As fusion occurs (third picture in the sequence) the cell walls and membranes around initially distinct tubes coalesce, so that their intracellular cytoplasm, which in its turn contains membrane bound organelles (nuclei and mitochondria) becomes continuous. A visible recoil can occur in the receptive hypha when the tubes coalesce (photographed by Dr A.M. Ainsworth)

labyrinthine branching systems and networks (i.e. as shown in Fig. 4.5). It is *not* due to the 'ties that bind all into a web of one' (Rayner 2004; Tesson 2006; cf. Barabási 2002).

4.5 Self-identity and Consciousness as a Natural Inclusion of Neighbourhood

This shift from regarding space and boundaries as sources of discontinuity between material bodies to understand them naturally as sources of continuity and energetic distinction, opens up our understanding of self-identity. A 'living I' cannot be a hermetically sealed, autonomous unit isolated from its neighbourhood, because the space within its distinctive but not definitive bodily boundaries is continuous with the space beyond these boundaries. It finds identity not in its inner self, alone, but in its variably receptive, reflective and responsive energetic relationship with its limitless and changeable surroundings. It is a natural inclusional 'I', not an abstract 'I'.

This distinction between the natural inclusional and abstract 'I' may correspond with the distinction made by Winnicott (1965) between the 'true self', which alone can feel real and be creative, and 'false self', which plays a protective but potentially pathological role. The ability to distinguish, but not necessarily define unique identities is a vital condition for intervention and participation in the world (Rayner and Jarvilehto 2008). A newborn baby may have no such sense of distinction between self and world, so that all that happens seems to happen to itself. The experience of meditative trance and what some have called 'no-self', 'core consciousness' and 'inspiration phase' mental activity (Harding 2000; Damasio 2000; Claxton 2006) may correspond with this lack of distinction and openness to all possibility. With the development of co-creative relationships with other people and outside world, however, the child needs to make distinctions between her/his body and others in order to receive, respond and provide directional guidance. An objective/subjective 'self-consciousness', 'extended consciousness' and 'elaboration phase' mental activity develops (Harding 2000; Damasio 2000; Claxton 2006), along with an awareness of personal joy and pain through learning experience of self-inclusion in natural neighbourhood. As this takes hold—and may even be regarded as a 'superior' form of 'intelligence' (Damasio 2000; Claxton 2006; cf. Harding 2000) it may, however, harden into objective definition.

By acknowledging ourselves as distinct but not isolated local inclusions of natural energy flow, it is always possible gracefully to accept what we receive, to nurture and make the best of it, eventually to pass it on. Such is the way of cultures that operate the co-creative relay of a gift economy (Hyde 2006; see Chap. 5). But trouble starts as soon as it seems possible to define and own what is morally or functionally best and remove or exclude what does not pass muster. To make such judgements it would be necessary to step *completely* outside the flow of what we are inescapably immersed in, in order to take a 'God's eye view'—or, in Darwinian terms, the view of a 'natural selector'. This is not possible, but when we nonetheless attempt to do it, as observers distanced from what we observe, we risk converting the true empathy and co-creativity that comes from sensing the *needfulness* that comes with being a receptive centre of energy flow into psychological projections of narcissistic self-reference (selfishness) and dependency (neediness) (cf. Neuman

2010). What may appear superficially to be good for the persistence of the individual or group from a definitive perspective may not be good for the sustainable flourishing (well-being and well-becoming) of all in natural, co-creative communion (Rayner 2011a).

With this understanding comes the possibility of removing the unrealistic grounds for opposition between 'each' and 'other(s)' that contribute to profound human conflict and environmental damage. Each individual finds identity not in the inner self, alone, but in the variably receptive, reflective and responsive energetic relationship with its changeable surroundings. This fundamentally psychological understanding holds the hope, perhaps the only hope, for sustaining the flourishing of humanity in a world that has been drawn to the brink of environmental and social breakdown through the assumption that space can be cut. As Polyani (1958) put it:

> For once men have been made to realize the crippling mutilations imposed by an objectivist framework – once the veil of ambiguities covering up these mutilations has been definitely dissolved – many fresh minds will turn to the task of reinterpreting the world as it is, and as it then once more will be seen to be.

4.6 Genes as Reproducers of Life Patterns, not Units of Selection

So, if genes are not discrete units of selection, what real role do they play in the emergence of self-identity and evolution by natural inclusion? Certainly they do have a crucial role. But they are not in the driving seat of individual character or evolutionary change, nor are they the subjects of an external selective force that judges their fitness for purpose. Rather they are deeply involved participants in an evolutionary process that occurs from sub-atomic to galactic scales of existence. This process originates in the mutual relationship between natural space and energetic flux from which all the varied but recurrent patterns of life I describe in this book emerge. In saying this, you may recognize that I am not confining what I mean by 'life' solely to what is generally regarded as 'biological life'. I see biological life not as an exception from an otherwise inanimate material existence, but as a special case of the 'life of the cosmos' that manifests from the inclusive relationship between receptive spatial stillness and informative flux in all local phenomena.

Yes, biological life is a special case in which DNA fulfils a role analogous to that of the grooves of a vinyl record that not only enable the same sound patterns to be reproduced and amplified again and again, but can themselves be reproduced again and again in pressing after pressing. These grooves are receptive configurations of energetic information and space that enable particular configurations of energetic information to be reiterated and reiterated. In the same way, the configurations of base-pairs within the receptive groove of a DNA spiral are transcribed into the genetic code of messenger-RNA that is translated via transfer-RNA into the alignment of the twenty kinds of amino acids from which protein is synthesized.

The spatial configuration of proteins then serves both structural and metabolic roles in the life of living cells, which affects the way they take shape and respond to their neighbours and neighbourhood as they take in the energy from their surroundings that they need to be able to function and reproduce. And, of course the recording within the genes can not only be reproduced, it can be altered and recombined in myriad ways, just as the way its message can be affected by the environmental circumstances it occurs within. This is the fundamental nature of heredity, and it by no means arises from the selfish, solitary, self-contained existence of genes as independent units of digital information that it is sometimes made out to be (Dawkins 1995).

4.7   Life, Love and Natural Inclusion

A scientific man ought to have no wishes, no affections, - a mere **heart of stone**.
Charles Darwin

The above statement not only tells us something about the effect abstract rationality has had on the way science is mostly practiced, but also how the depiction of biological evolution as a ruthless battleground between selfish genes striving for individual supremacy could ever have captured human imagination. The deliberate exclusion of what is regarded as emotional subjectivity in order to be 'objective' actually has the effect of introducing extraordinary bias into scientific praxis. This praxis eschews from consideration the three occurrences fundamental to comprehending natural inclusion:

- Receptive space
- Informative flux
- The co-creative inclusion of each in the other.

In so doing, life is rendered into a lifeless, loveless set of inert material units of selection struggling to survive in the face of intolerant external forces.

For what those three aspects of natural inclusion represent most fundamentally are nothing less than the three 'Aspects of Love' that have been recognized by some spiritual traditions.

Agape: the unconditional, welcoming, nurturing, receptive acceptance of life in all its guises by 'World **Soul**'.

Eros: the responsive, passionate, living **Spirit** of natural energy flow within and between all natural forms.

Philia: the **Bodily** companionship of each within the other's compassionate reach.

They also signify the deep, abiding relationship between Female, Male and Offspring.

or
Mother, Father and Child.
And the Yin, Yang and each in the other of Daoism.
The Living Light of Passion in the Loving Darkness of Compassion.

The ascendency of abstract thought that arose during the 'Scientific Revolution' divorced 'Reason' from 'Emotion'. Awareness of natural inclusion reveals why reason cannot be separated from emotion, because emotion is a manifestation of natural energy flow.

So, it might well be said that natural inclusion restores the love, life and art lost by abstract thought, by offering the kind of soulful and spiritual enquiry that befits truly 'natural' science!

References

Ainsworth, A. M., & Rayner, A. D. M. (1986). Responses of living hyphae associated with self and non-self fusions in the Basidiomycete *Phanerochaete velutina*. *Journal of General Microbiology, 132,* 191–201.

Barabási, A-L. (2002). *Linked: The new science of networks*. New York: Perseus Publishing.

Bateson, G. (1972). *Steps to an ecology of mind: Collected essays in anthropology, psychiatry, evolution, and epistemology*. University Of Chicago Press.

Bertalanffy, L. (1968). *General system theory: Foundation, development, applications s.* New York: George Braziller.

Cairns, H.C. & Harney, B.Y. (2004). *Dark sparklers*. H.C.Cairns

Claxton, G. (2006). *The wayward mind*. London: Abacus.

Damasio, A. (2000). *The feeling of what happens: Body, emotion and the making of consciousness*. London: Vintage.

Darwin, C. (1859). *On the origin of species by means of natural selection, or the preservation of favoured races in the struggle for life*. Bromley, Kent: Down.

Dawkins, R. (1989). *The selfish gene* (New ed.). Oxford: Oxford University Press.

Dawkins, R. (1995). *River out of Eden: A Darwinian view of life*. USA: Basic Books.

Dowson, C. G., Rayner, A. D. M., & Boddy, L. (1986). Outgrowth patterns of mycelial cord-forming basidiomycetes from and between woody resource units in soil. *Journal of General Microbiology, 132,* 203–211.

Elstrup, O. (2009). The ways of humans: Modelling the fundamentals of psychology and social relations. *Integrative Psychological and Behavioural Science, 43,* 267–300.

Elstrup, O. (2010). The ways of humans: The emergence of sense and common sense through language production. *Integrative Psychological and Behavioural Science, 44,* 82–95.

Gabriel, Y. (2002). Essai: On paragrammatic uses of organizational theory—A provocation. *Organization Studies, 23,* 133–151.

Harding, D. E. (2000). *On having no head—Zen and the rediscovery of the obvious*. London: The Shollond Trust.

Heelas, P., & Lock, A. (1981). *Indigenous psychologies: The anthropology of the self*. London: Academic Press.

Hutchison, C.A. (2016) Design and synthesis of a minimal bacterial genome. *Science 351*(6280).

Hyde, L. (2006). *The gift—How the creative spirit transforms the world*. Edinburgh: Canongate Books.

Koestler, A. (1976). *The ghost in the machine*. London: Hutchinson.

Lewis, C.S. (1942). *The screwtape letters*. Oxford: Geoffrey Bles.

Marman, D. (2016). *Lenses of perception—A surprising new look at the origin of life, the laws of nature and our universe*. Ridgefield, Washington: Lenses of Perception Press.

Marman, D. & Rayner, A. (2016). The littlest genome and the question of life. https://www.bestthinking.com/articles/science/biology_and_nature/genetics_and_molecular_biology/the-littlest-genome-and-the-question-of-life

Neuman, Y. (2010). Empathy: From mind-reading to reading of a distant text. *Integrative Psychological and Behavioural Science*. doi:10.1007/s12124-010-9118-7.

Polanyi, M. (1958). *Personal knowledge: Towards a post-critical philosophy* (p. 381). London: Routledge and Kegan Paul.

Rayner, A. D. M. (1997). *Degrees of freedom—Living in dynamic boundaries*. London: Imperial College Press.

Rayner, A. D. M. (1998). Fountains of the forest—The interconnectedness between trees and fungi. *Mycological Research, 102*, 1441–1449.

Rayner, A. D. M. (2004). Inclusionality and the role of place, space and dynamic boundaries in evolutionary processes. *Philosophica, 73*, 51–70.

Rayner, A. D. M. (2010). Inclusionality and sustainability—attuning with the currency of natural energy flow and how this contrasts with abstract economic rationality. *Environmental Economics, 1*, 98–108.

Rayner, A. D. (2011a). Space cannot be cut: Why self-identity naturally includes neighbourhood. *Integrative Psychological and Behavioural Science, 45*, 161–184.

Rayner, A.D.M. (2011b). *NaturesScope: Unlocking our natural empathy and creativity—an inspiring new way of relating to our natural origins and one another through natural inclusion*. O Books.

Rayner, A. D. M., & Jarvilehto, T. (2008). From dichotomy to inclusionality: A transformational understanding of organism-environment relationships and the evolution of human consciousness. *Transfigural Mathematics, 1*(2), 67–82.

Sampson, E. (1988). Indigenous psychologies of the individual and their role in personal and societal functioning. *American Psychologist, 43*, 15–22.

Sampson, E. (2000). Reinterpreting Individualism and collectivism: Their religious roots and monologic versus dialogic person-other relationship. *American Psychologist, 55*, 1425–1432.

Shakunle, L.O. & Rayner, A.D.M. (2008). Superchannel—Inside and beyond superstring: the natural inclusion of one in all—III. *Transfigural Mathematics, 1*(3), 9–55, 59–69.

Shakunle, L. O., & Rayner, A. D. M. (2009). Transfigural foundations for a new physics of natural diversity—The variable inclusion of gravitational space in electromagnetic flow-form. *Journal of Transfigural Mathematics, 1*(2), 109–122.

Smith, B. (1997). Boundaries: An essay in Mereotopology. In L. Hahn (Ed.), *The philosophy of Roderick Chisholm* (pp. 534–561). La Salle: Open Court.

Taylor, S. (2005). *The fall*. Winchester, New York: O Books.

Tesson, K.J.A. (2006). Dynamic networks: an interdisciplinary study of network organization in biological and human organizations. Ph.D. Thesis, University of Bath.

Valsiner, J. (2009). Baldwin's quest: A universal logic of development. In J. W. Clegg (Ed.), *The observation of human systems—Lessons from the history of anti-reductionist empirical psychology* (pp. 45–82). New Brunswick, London: Transaction Publishers.

Walker, E.M. (2003). The confusion of dreams between selves and the other: Non-linear continuities in the social dreaming experience. In W.G. Lawrence (Ed.), *Experiences in social dreaming* (pp. 215–227). London: Karnac Books.

Wilber, K. (1996). *A brief history of everything*. Boston: Shambhala Publications.

Wilson, E. O. (1998). *Consilience—The unity of knowledge*. London: Little, Brown and Company.

Winnicott, D. W. (1965). Maturational Processes and the facilitating environment. London: Hogarth Press.

Chapter 5
Flow Geometry and the Evolution of Collective Organization: Individuals, Couples, Series and Huddles

Abstract Extraordinarily varied patterns of structure, function and relationship emerge and co-evolve over nested scales of organization in living systems. These patterns recur in cells, individuals, species and ecosystems. Their evolution depends on the generation and sustainability of diversity, not its progressive elimination in favour of what is specially privileged or advantaged. It truly does 'take all kinds to make a world', whether that 'world' is an individual cell, an assemblage of cells, an assemblage of individuals, an assemblage of populations, an assemblage of communities or the limitless space of Nature everywhere, which includes all distinguishable localities of place and time. This diversity ultimately arises simply and intrinsically from the *relationship* between receptive space and informative flux as source of all distinguishable natural bodily forms, and how these forms in turn relate to one another. In biological systems, it emerges within the receptive medium of water. Organisms are hence better understood as embodied water flows in mutual relationship with one another than as genetically controlled machines. Natural diversity comes without the intervention of any central or external administrative agency. It is orderly but not rigidly ordered, being neither random nor preordained, but fluid.

5.1 Natural Companionship

> Nowhere is an island, entire of it self

Which is truer to Nature and human nature: individualistic competition or collectivistic co-operation? That is the central question underlying the opposition between two alternative abstract views of evolution and psychology that persist to this day. At its heart is the assumed independence of individual or group as a singular entity complete in itself. The inherent paradox and potential for enmity and extremism embedded in this assumption is evident in two commonly expressed attitudes: 'either you're with me or against me', the individual opines to others not defined as 'self'; 'either you're with us or against us', the group opines to outsiders not included in its membership. No other possibility is considered acceptable. Borders

© The Author(s) 2017
A. Rayner, *The Origin of Life Patterns*, SpringerBriefs in Psychology and Cultural Developmental Science, DOI 10.1007/978-3-319-54606-3_5

between 'in' or 'out', 'pro' and 'con' are absolute, and defended with utter intransigence.

Close your eyes and try to imagine yourself entirely alone. What do you feel? Exultation, grief, peace, anxiety, freedom, exposure, security, confinement, hope, despair, ecstasy, agony, excitement, boredom, warmth, coldness, anything else, nothing at all? Can you imagine that life would be possible or sustainable in this situation, and if so, how?

Now, imagine that you have company, but that this is in every respect identical to yourself, like an army of model soldiers, each placed an equal distance apart from every other in an absolute, endless, uniform, ordered array. What do you now feel? How long could this situation be sustained, given that the movement out of position of any one soldier will change the outlook of each and all soldiers, especially his/her/its nearest neighbours? What would be needed to sustain this uniformity?

Now imagine that you have company, but that this company is variable, both human and non-human, and is capable of movement, even though it may have at least some features in common with yourself. This should not be too difficult to imagine, because it coincides with your everyday experience. But how do you feel about your neighbours and neighbourhood, and how do you relate to them?

As this book reveals, we are *both* individually unique *and* dependent on relationships with others to sustain ourselves. In fact it is *because* of our differences, and our associated *complementary* strengths and weaknesses, not in spite of them, that our need for relationship exists. Were we *actually* all identical, independent whole units, complete in ourselves, we could all lead entirely separate existences without any kind of interference or coherence whatsoever. But we are not and we cannot. Instead we are living, breathing, feeding, excreting, transient creatures who have no choice but to live in communities with others, both human and non-human.

We are neither all alone, as free agents, nor all parts of one and the same seamless entity. We are distinct, but not isolated from one another. We can only live in one another's company, and the kind of company we keep hugely influences our quality of life. How could we possibly think otherwise?

The fact that we can and do often think otherwise is due to the abstract perception that *dislocates* material entities from the space they naturally include and are included in, setting them in opposition to each other. This forces us to conclude, falsely, that things are either isolated from one another by space as definitively bounded whole objects, or that they are parts of one and the same whole object. It prevents us from understanding and behaving in accordance with our natural companionship as dynamic inhabitants of our natural neighbourhood. We envisage ourselves as separate from and indeed superior to other forms of life and project upon these our own demonised natural psyche, which Freud called the 'Id' and Jung called the 'Shadow.' In popular culture, when we regard other humans as behaving badly, we call them 'animals', 'monsters' and 'savages', and when we regard other humans as mentally inadequate, we call them 'vegetables'. This denigration of non-human life is sustained by Darwinian and neo-Darwinian depictions of biological life as inescapably 'selfish' (see Chap. 4). Meanwhile, many of us who consider ourselves to be 'civilised' are prepared nonetheless to condone the extreme violence of war as well as to accept and reinforce the vast disparities in wealth and quality of life that are all-too-evident in modern cultures.

The very idea that humanity has anything in common with or can learn from non-human life has widely been regarded either as an insult or as an excuse for competitive, self-serving behaviour (cf. Dawkins 1989; Chap. 4). But if we can truly accept and appreciate our human evolutionary origin in natural communion with other forms of life, then I think we can gain insights from this other life into our own ways of living. This is my purpose in including this chapter as an overview of scientific knowledge about recurrent patterns of natural diversity in non-human living systems, from cellular to global, and how these arise from co-creative relationships between distinct but not independent identities. In the following sections, it may be helpful to refer back to Figs. 2.12 and 2.13, which illustrate many of the features I will be describing.

I will begin by paying attention to that vital ingredient that is so often overlooked in discussions of the evolutionary origin of biodiversity.

5.2 Biological Life as an Embodied Water Flow, not a Genetically Controlled Machine

The idea that living organisms are sophisticated machines has long been with us as a product of mechanistic thinking. Before evolutionary theory became generally accepted, organisms were widely regarded as the creations of an external maker, in much the same way that we regard the products of our own manufacturing industry. To find out more about these creations, techniques of dissection and vivisection were used to explore the anatomy below the outward appearance of organisms and identify the functions of different parts. In taking the 'living machine' apart in this way, the hope was that we might not only be able to understand its workings but also repair and even re-assemble or re-create it when it went wrong, expired or needed improvement. This hope lives on in the modern development of genetic engineering, transplant surgery, infertility treatment, cloning and regenerative medicine. Yet to many minds it is profoundly dangerous to our psychological and social well-being.

There is something deeply troubling about regarding human and other organisms as reproducible living machines, which Mary Shelley's depiction of Frankenstein's Monster was intended to evoke. Some *vital quality* is missing from this mechanistic picture. The Romantic Movement sought to point out this missing quality. But it was overlooked in the rush for technological progress. Meanwhile, far from using an understanding of evolutionary process to expose the myth of the externally created machine-organism, adherents of natural selection theory developed an even more emotionally alienating metaphor.

During the twentieth century, the widespread acceptance of Darwinian theory combined with the discoveries of genetics and development of information technology to convert the natural organism into a fully autonomous form of artificial intelligence, complete with hardware and software. The organism became a set of designer genes in a robotic body whose only function was to convey digital information from one generation to the next as exclusively and prolifically as possible (Dawkins 1995). This selfish replicator came fully equipped to exterminate its opposition and calculate the costs and benefits of its actions in order to make more of itself in what, apart from random accidents, is a fully definable future.

Current evolutionary theory is correspondingly largely a field of sophisticated gamesmanship, using more and more complex digital mathematical models to work out how one gene, or set of genes can outdo another in a specific arena of conflict. What could be missing from this picture?

When scientists explore the possibility of biological life evolving on other planets, the first substance they look for is not DNA, but an altogether much simpler chemical compound: water. Why? What is it about water that is so *vital* to biological evolution?

Quite simply, in carbon-based biological life, water takes on the role of *receptive medium* that space has universally. Remember my description in Chap. 2 of the need to add solvent to paint pigment in order to create a painting? Life, when active, is wet, not dry! Underlying this role is the relational chemistry of what happens when electrons flow around and betwixt atomic centres.

Water is a surface-tense *liquid* at atmospheric pressure and between 273 and 373 K (freezing 'point' and boiling 'point'). This is amazing in view of its low molecular weight, but can be understood in terms of the receptive–responsive relationship of oxygen and hydrogen atoms, which gives rise to a phenomenon known as 'hydrogen-bonding'. The inductive space within an oxygen atom is *receptive* to what is conventionally labelled 'negative charge', conveyed by electrons, whereas a hydrogen atom *responsively releases* negative charge from the environs of the 'positively charged' proton in its nucleus.

Correspondingly, water molecules are *dipolar*, with electrical charge being distributed unevenly so as to be relatively negative in the vicinity of oxygen and relatively positive in the vicinity of the two hydrogen atoms. Complementary charges attract, like male and female, and so water molecules *couple* with one another to variable degrees to form complex associations. These associations become more fluid as temperature increases towards boiling point, and stiffer as temperature decreases towards freezing point. Significantly, the densest configurations occur 4 K *above* freezing point, so that ice floats on water—and life can continue to move around in seas, lakes, ponds and puddles whose surface is frozen.

This electrical dipolarity is also very significant in enabling water to participate in the dynamic relational flow-form of organisms as a *differential solvent*—a medium that dissolves and so fluidises materials to varying degrees depending on their composition. 'Polar' materials are 'hydrophilic' ('water-loving') and dissolve readily. These include salts, sugars, acids, alkalis, alcohol and *some* parts of proteins. 'Non-polar' (neutrally charged) materials are 'hydrophobic' (water-repellent), and do not dissolve readily. They include fats, oils and *other* parts of proteins. Whereas hydrophilic materials are permeable to water flow, hydrophobic materials are impermeable.

By bringing together a suitable mix of hydrophilic and hydrophobic materials in a watery context, it is possible for living organisms to assume an enormous variety of flow-forms, based on the development of what is known as 'osmotic pressure' within a partially permeable bodily boundary. This variety corresponds in many ways with the forms that water itself assumes within, upon and above Earth's surface. Bodies of water can dissociate into variably sized individual droplets and branches and associate into streams, rivers, pools, clouds and labyrinthine networks, depending on how and where they gather in, distribute and conserve sources of energy in diverse circumstances. The same kinds of river-like pattern can be found in slime bacteria swarms, slime mould plasmodia, fungal colonies, army ant swarms, wildebeest herds, trees, leaf venation, blood systems and nervous systems. The same kinds of currents, ripples and vortices as those found in ocean and atmosphere occur both within individual bodies and collective groupings of organisms ranging in scale from minuscule to global.

All in all, life as an embodied water flow is a far cry from the suspended animation of dormant life. The latter conserves energy in seeds, spores, eggs, cysts, sclerotia, pupae, etc. It does not 'compete to survive', but 'survives to revive'.

How these embodied water flows of life thrive and survive in varied guises as they encounter different circumstances in their real-world context depends upon the hydrophilic and hydrophobic characteristics of their enveloping boundaries. These boundaries are not discrete limits, but instead dynamic interfacings through which their insides and outsides are both distinguished and reciprocally coupled. Their characteristics affect the degree of *openness* and corresponding *resistance* to *expansion*, *uptake* or *leakage* from or to outside and *internal fluid flow*. In other words, they influence boundary *deformability*, *permeability* and *connectivity*, as illustrated in Fig. 4.2.

So, let us now take a more detailed look at the diversity of life as an embodied water flow, and how this diversity exemplifies the natural communion of each in the other's receptive and responsive influence at different scales of biological organization. Much of what I am about to summarize can be found in any standard biological textbook: what is important here is to appreciate how this relates to the basic theme of natural inclusion.

5.3   Diversity Within Cellular Life

As mentioned and illustrated in Chap. 2, modern microscopic and biochemical techniques have revealed the internal structure and functioning of living cells to be far more elaborate, fluid and complex than was expected when these entities were first discovered and named. Crucial to cellular organization is the presence of a watery medium, protoplasm, enveloped within a dynamic envelope, which incorporates both its store of genetic information and the cytoplasm in which the energetic processes required to sustain life occur. The dynamic envelope mediates the cell's vital relationship with its local environment. Both the bounding of the cytoplasm and the energetic processes within it involve a complex relationship between water-soluble and water-insoluble compounds including proteins, lipids and carbohydrates.

In the 'prokaryotic' cells of bacteria there is usually a cell wall, which confers structural rigidity, and a plasma membrane enveloping the cytoplasm. But there is no internal partitioning between the genetic material and cytoplasm, or between different cytoplasmic components. There is hence little scope for functional differentiation between different localities within the cell's interior.

In the 'eukaryotic' cells of fungi, plants and animals, there is by contrast considerable internal differentiation into structurally and functionally distinctive localities. DNA is contained in bodies called chromosomes, and these are contained within a true nucleus that is partitioned from the cytoplasm by a double layer of membranes with pores in it. There are also membrane-bound organelles such as mitochondria, chloroplasts (in photosynthetic cells), Golgi bodies, lysosomes and peroxisomes. There may be one or more solution-filled vacuoles, and there are membrane networks known as endoplasmic reticulum, which ramify through the cytoplasm. Ribosomes are assembled in a distinct region of the nucleus, the nucleolus, and are exported into the cytoplasm where they either remain free in the cell sap ('cytosol') or attach to the endoplasmic reticulum. Within the cytosol may be cytoskeletal proteins, which provide for support and movement within the cell.

The presence of DNA and ribosomes within chloroplasts and mitochondria has been taken to imply that these may originally have been free-living prokaryotes that became integrated by enclosure within another ancestral cell as an 'endosymbiosis' (see below). Eukaryotic cells are therefore veritable packages of packages, dynamically bounded and internally partitioned by fluid membranes (Rayner 1997). By their very nature, living, active cells cannot be absolutely independent sealed units, any more than can their contents. Rather, they thrive in the dynamic relationship between their insides and outsides, mediated through their variably permeable, deformable and connective envelopes. It therefore makes more sense to regard them as receptive-responsive fluid configurations of space than as discrete object-boxes.

5.4 Diversity Within Multicellular Life

The significance of viewing cells in fluid neighbourhood becomes apparent as soon as we consider the relationships of two or more cells co-existing alongside one another to form couples, linear series and huddles.

To sustain its biochemical activity (metabolism), a cell uses chemical energy to pump molecules and ions through channels in its semi-permeable membrane. This is analogous to a petrol pump that uses petrol to fill itself with more petrol. Suppose two such pumps occur next to one another. The more active pump could drain its neighbour of nutrients. This actually happens in some parasitic and cancerous conditions. It can be avoided by the opening up of communication channels between adjacent cells—and such channels are a vital feature of multicellular organization in plants and animals. The cells are in natural communion with one another, not sealed boxes.

An even more salient illustration of the limitations of the 'sealed box' portrayal of the cell can be found in many fungi and fungus-like organisms and parts of organisms. Here, cell growth is characteristically confined to a parabolic dome-shaped tip. Elongation of the cell boundary from this tip gives rise to a tube, known as a 'hypha'. This tube can continue to elongate, as well as producing more elongating tubes by branching, as long as it is supplied with water and nutrients, to form a collective system known as a 'mycelium'. Moreover, the tips of the branches within this system can fuse (anastomose) with one another (see Chap. 4, Fig. 4.5), so converting the initially 'dendritic' (tree-like) system of radiating branches into a dynamic, labyrinthine network. Whilst organelles may well be partitioned into a *series* of distinct cellular compartments along the length of the tubes by cross-walls called 'septa', internal communication is sustained through gaps in these walls.

The bodies of many plants and animals contain distinct arrays or *huddles* of cells called 'tissues'. These tissues may also be contained within 'organs' that are 'specialized' to serve different, complementary functions in the life of the organism. That this complex organization arises endogenously, in situ, not exogenously, via the assembly from elsewhere of discrete, pre-existing cellular units is immediately obvious from observations of embryonic development. Tissues and organs develop through processes of cellular proliferation and individuation from the living protoplasm of a relatively large and immobile receptive female 'egg' cell, after 'fertilization' by a relatively tiny and highly mobile 'sperm' cell. This coupling of egg and sperm recalls the complementarity of 'female' receptive space and 'male' responsive flux that ultimately gives rise to all distinguishable natural bodily form.

In many ways, embryonic development beautifully expresses the transformation of evolutionary flow-form. At first, after the fertilized egg or 'zygote' has multiplied by dividing internally a few times, most embryos consist of little more than a huddle of more or less similar cells. For tissues to form, some kind of re-organization has to occur, so that the cells become distributed into distinctive layers or regions where they follow different developmental pathways. The way this re-organization occurs contrasts markedly between most plants and animals, reflecting the difference between those forms of life that *grow* from place to place and those that *move* bodily from place to place.

In plants, the embryo transforms into a cylindrical structure. This structure grows in an analogous way to fungal hyphae: new cells form at its tips of via division from either a single apical cell, or a group of cells known as a 'meristem'. Further tips may then arise by means of branching.

In most animal embryos, the production of new cells occurs within all the developing organs and tissues. Here, from the viewpoint of an external observer, development appears to be highly *prescriptive*, occurring in a set sequence and directed towards a specific end point—the sexually mature adult. Viewed from within the developing body boundary, however, indeterminate processes are evident, and even minute variations in the relational dynamics of these processes can have radically different outcomes.

Following fertilization of an animal egg, the number of cells and/or nuclei increases via a series of doublings. As this process continues, an internal hollow or 'blastocoel' develops, preparatory to a remarkable phase of boundary-infolding, known as 'gastrulation', which culminates in the formation of inner, outer and intermediary tissue layers. Cells within these layers ultimately become specialized for distinctive roles in skin, nerve, gut, muscle, connective tissue, bone, blood vessels, liver, kidneys, etc.

The processes that follow gastrulation are generally considered to involve an 'epigenetic' programme that activates and inactivates the *expression* of distinctive sets of genes. Progression through this programme through a process known as *determination* equips the emerging adult for engagement with its habitat. Each move to specialization is conditional upon those paths that have already been followed and with a few exceptions, determined animal cells cannot change their developmental course—only so-called stem cells retain this ability. By contrast, even fully differentiated plant cells can regenerate into adult form. As a result, plants are more able to change their pattern of *development* in response to environmental change, whereas animals can change only their pattern of *behaviour*.

The separation of distinctive life-maintaining functions into local regions or tissues with specialized attributes allows each efficiently to get on with its own work, with minimum interference from others. But this specialization also results in an inability to function in isolation and so some kind of *communication* system that interconnects their activities is needed. This generally consists of pipelines that either conduct fluid, as in the 'vascular' systems and air channels of plants and animals, or electricity, as in the nervous systems of animals. The pattern of development of these pipelines is fundamentally river-like as they connect up their sites of supply and distribution.

5.5 Diversity Within Social Life

Similar processes of integration and differentiation to those in cellular and multicellular development also occur within and between groups of organisms. As organisms proliferate in energy-rich situations, they dissociate into branches and individual bodies with a large exposed surface that is unsustainable without continual replenishment of resources. As the external availability of resources diminishes, however, more coherent and persistent organizations develop *exogenously* due to boundary integration between initially distinct individuals (see Chap. 4, Figs. 4.2, 4.3, 4.4 and 4.5). At least initially, these organizations are more loosely held together than those arising endogenously, but their association may strengthen in time.

The interplay between specialization and communication makes absolute definition of social structure impossible. But this has not prevented 'individual organisms' from being treated as the fully definable 'building blocks' from which fully definable collective units are assembled.

In conventional ecology, a population is defined as a collection of individuals of the same species, but this begs the question of what individuals are, as well as where a collection of them begins and ends and what a species is.

As described in Chap. 3, idealized mathematical models assume that populations consist of discrete individual units in a uniform field. These models do not recognize that we Earth-dwellers actually live in the highly heterogeneous surface of a sphere with no fixed corners. This surface includes organisms, like fungi, and social formations like those of ants and locusts whose indeterminate growth potential may be restrained only by availability of resources. Any realistic model of populations therefore needs to include spatial and temporal heterogeneity both in the form of the population members themselves and in the scale and distribution of the places in which these members locate and assimilate resources.

Genetic variation becomes important when considering the implications of encounters between neighbouring population members. How does genetic difference arise and how does it affect our ideas about what constitutes a 'species'.

Throughout biological life on Earth, two alternative responses are evident in encounters between genetically different population members: a mutual or one-sided warding off or 'rejection' and a mutual or one-sided embrace or 'acceptance'. Rejection or 'somatic incompatibility' generally appears in such guises as territoriality and immunity, which enable the 'body' or 'soma' to maintain its identity by keeping itself to itself. Acceptance is 'sexual', resulting in the formation of a couple, which is often distinguishable as male and female identities.

It seems that the basic tendency for genetically unalike bodies to dissociate from one another is overridden by sexual attraction. If the attraction is insufficient, the dissociation will prevail and the encounter will be sexually 'incompatible'. Sexual incompatibility due to genetic difference gives rise to new species.

At first sight, incompatibility, and the seemingly aggressive responses that can accompany it may appear to be an expression of *opposition* to other and resultant *competition* and *conflict*. I think it makes better sense, however, to regard it more in terms of dissonant flow-patterns. I became aware of this possibility when I noticed that the interference patterns formed between adjacent ripples of water appear remarkably similar to those formed by fungal colonies growing in juxtaposed culture. The margins of the colonies surge out like a wave front. Quite often, rhythmically alternating, concentric ridges and troughs of aerial and submersed mycelium develop. Upon meeting, a trough or a ridge forms at the interface of the colonies. Where the colonies are genetically identical, and the peaks and troughs correspondingly of equal frequency and amplitude, this initial interfacial distinction often disappears as the aerial and submersed zones align and merge harmonically with one another. Where the colonies are not genetically identical, the interface between them either persists and intensifies as a mutual 'barrage' zone or is superseded by the emergence and spread in one or both directions, following mating, of a new or 'secondary' mycelial phase. Examples are shown in Figs. 5.1 and 5.2.

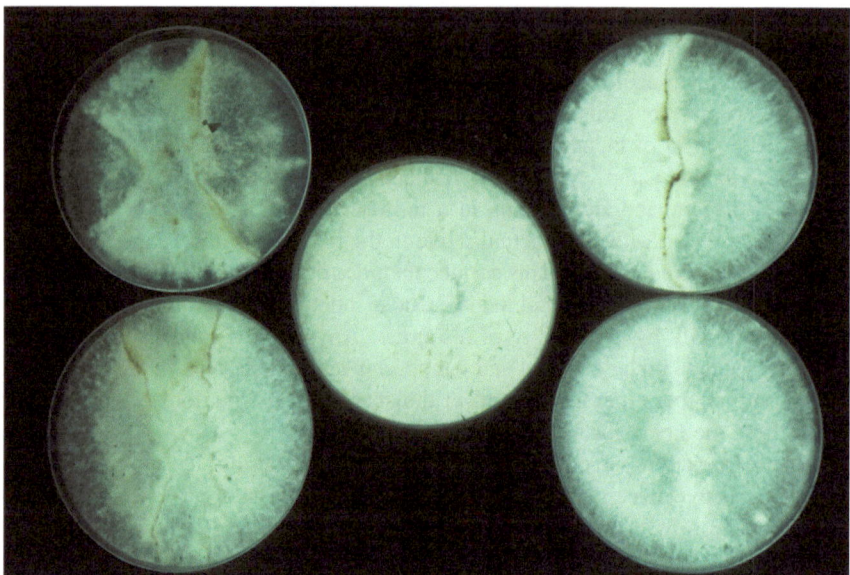

Fig. 5.1 Encounters between mycelia of the basidiomycete, *Stereum hirsutum*, inoculated side by side in Petri dish cultures. The encounters on the *right* were incompatible, that in the centre was compatible and those on the left were initially compatible but superseded by an incompatible response (photograph by A.M. Ainsworth)

My mention of fungal colonies raises another question. What is a colony? Colonies are usually defined as associations of unicellular or multicellular organisms of the same species. How does this differ from a population or society? Conventionally, members of a colony are separable yet live closely together in a particular locality. But how separable, how close together and how localised? How alike in form and behaviour are members of a colony? When are they so functionally specialized as to be more aptly regarded as members of a society?

Simple gatherings of similar organisms, such as herds, flocks and shoals of animals or clumps, tufts and stands of plants occur whenever the organisms are contained within a physical boundary, are mutually attracted in some way or do not detach fully when they multiply. The boundaries of these simple gatherings may be relatively sedentary or they may be sufficiently fluid to enable them to generate an immense variety of patterns by both creating and following paths of least resistance.

As in embryonic development, initially simple arrays of similar forms can give rise to elaborate structures through diversification into different specialized roles linked together by communicative channels. Amongst animals, two examples of such complex social structures are found in certain jellyfish-like creatures: the 'hydroids' and 'siphonophores'.

Hydroids consist of individual 'polyps'—goblet-like forms whose gut cavities are all connected to one another, usually by a tubular system of erect branching 'stems' and creeping stolons or 'hydrorhiza'. The hydrorhiza extend outwards and

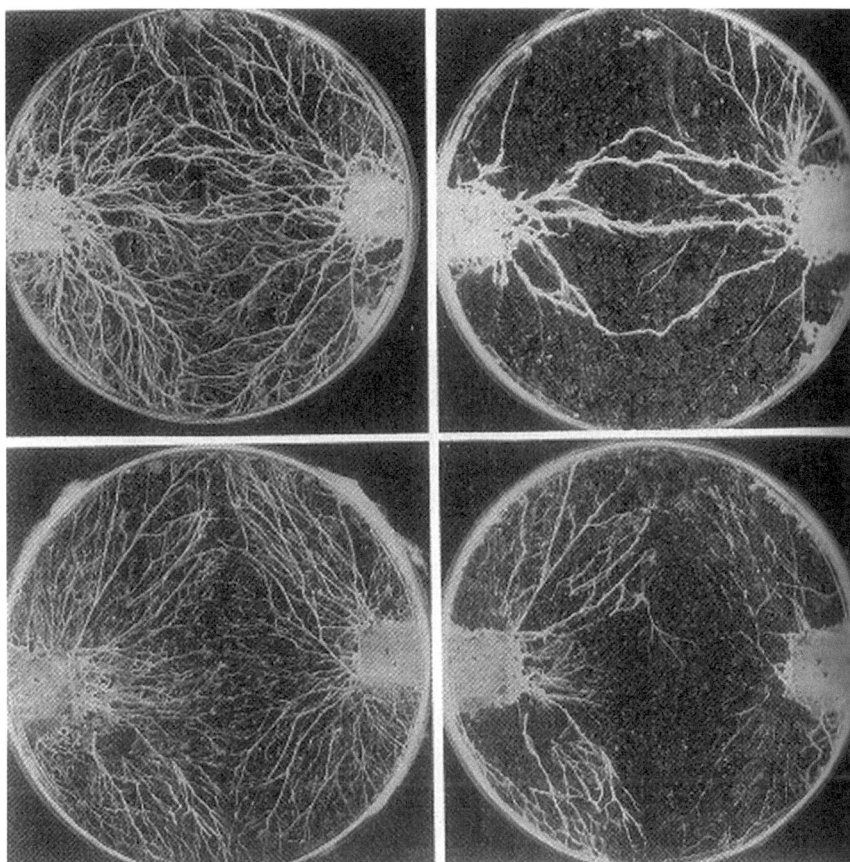

Fig. 5.2 Early (*left*) and late (*right*) stages in compatible (*upper*) and incompatible (*lower*) pairings between mycelia of the wood-decaying fungus, *Phanerochaete velutina* grown from wood blocks inoculated into soil, showing the formation of persistent and degenerating channels across the zone of overlap (from Dowson et al. 1989)

give rise to further erect stems, so increasing the size of the colony. The polyps develop in a variety of distinctive functional forms. Many are feeding or stinging polyps. Others give rise to free-swimming, bell-like 'medusae' that produce sexual offspring. Siphonophores are complex assemblies not only of different kinds of polyp, but also different kinds of medusae, all communicating with one another through internal channels. A well-known example is the Portuguese-man-of-war.

   Another kind of social organization, in which organisms work together through their behaviour rather than via direct bodily channels, is found in some kinds of insects, like ants and bees. Here, very different looking forms occur, known as 'castes', which have reproductive, feeding and protective roles, e.g. 'queens', 'workers' and 'soldiers'. Specialization into different social roles also occurs in prides of lions, troops of baboons, packs of hyaenas, societies of meerkats and human societies, although this is less strongly associated with differences in bodily appearance.

In social groupings that are pooled together through their behaviour rather than via direct bodily channels, members of the same collective are generally closely related—as members of the same family—but are not genetically and spatially identical in origin. Sexual outcrossing between members of one family and another family produces different families. Members of different families are therefore more genetically diverse than are members of the same immediate family.

The interplay between sexual and somatic relationships has important implications for human cultures. Colonialism and migration over large geographical distances have hugely disrupted our natural neighbourhood relationships, not only amongst ourselves, but also with the other life forms that have accompanied us on our travels. To understand more fully both the dangers and creative possibilities of this situation, we need to take into account a yet larger picture of the organization of life on Earth.

5.6 Diversity Within Community Life

The interplay between processes of specialization and communication is also crucial to the organization and evolution of natural community life. But what is a natural community? Again, the impossibility of absolute definition has not deterred many from regarding the species as the basic 'building block' from which the community as a 'whole' is assembled.

In conventional ecology, a natural community is defined as the living component of an ecosystem. In containing diverse species, it is not the same as what is commonly called a human community, which is more akin to what is described biologically as a colony or society.

Like human societies, natural communities are characterized by having at least some functional coherence, but, *tellingly*, this is *not* generally controlled by any obvious governmental structure or monarchy (see Chap. 6). Natural communities are 'self-orchestrating'—their coherence arises from the complementary form and functioning of their members pooled together in common space. For a natural community to continue to thrive, however, at least some of its members need to be 'primary producers', transforming sunlight or inorganic sources of energy into organic form. In many (but not all) situations exposed to sunlight, these primary producers are plants. Through their life and death, they form a dynamic framework within, upon and around which heterotrophic organisms, incapable of producing their own food, dwell in ever-changing natural communion.

Since they include identities assembling together exogenously, from disparate sources, the 'building block' metaphor does apply more readily to natural communities than it does to collectives emerging through proliferation from the same source. The basic principle underlying this assembly is the attraction of 'one' to the receptive space of an 'other', which one way or another serves as its 'host'. This receptive 'host space' may be in the form of another organism, whereupon it gives rise to what is known as 'symbiosis', or it may be in the form of some geographical feature.

The word 'symbiosis' was first used in the nineteenth century by Anton de Bary to mean a combination of two or more organisms living together. The most fully developed forms of symbiosis are generally regarded as 'mutualistic', where each organism benefits from, and indeed may be dependent for its viability upon the presence of the other.

In terrestrial ecosystems, the majority of multicellular plants would be unable to thrive without forming mutualistic partnerships, known as 'mycorrhizas', with fungi that enter and serve as absorptive accessories to their roots. The fungi extend out, as mycelium, into soil and thereby provide their plant partner with improved access to mineral nutrients and water in exchange for organic compounds produced by photosynthesis. The mycelium can also interconnect different plants—even of different species. By providing communication channels between the plants, mycorrhizal mycelia are thought to enable adult plants to nurture seedlings and to enhance efficient usage and distribution of soil nutrients. When we look at a forest or other stand of vegetation, we may be deceived by its superficial appearance into regarding it as an array of separate branching sticks in the ground that can do no more than bump into one another as they grow and sway in the breeze. But the reality underground, out of sight and out of mind, is that the plants are connected to varying degrees by complex, genetically diverse networks of fungi, like solar powered fountains linked together by hidden pipelines (see Chap. 4, Fig. 4.1).

Where larger plants are unable to establish in terrestrial habitats, another kind of symbiotic couple, lichens, covers surfaces that would otherwise be bare. Lichens consist of a photosynthetic filling of green algal or blue-green bacterial cells sandwiched between layers of fungal mycelium (a recent study has also indicated that this may be infiltrated by a yeast, resulting in a three-way partnership). Being tolerant of extremes of temperature and water availability, they grow very slowly, contributing over many years to processes of rock erosion and soil formation, and are a source of a unique variety of chemical compounds.

Not only terrestrial plants, but also many animals depend on mutualistic symbioses. The guts of many animals contain assemblages of microorganisms that both benefit from and can aid digestive processes. Some animals even cultivate partners that can aid digestion: amongst insects these include the wood wasps, ambrosia beetles, higher termites and attine ants, which grow 'fungus gardens'.

Mutualistic symbioses are also of great importance in marine communities. The reef-building corals, for example, depend on the presence of photosynthetic 'zooxanthellae' within their tissues and so cannot exist below depths where an adequate supply of light can penetrate.

Mutualistic symbioses have the potential to become so intimate that what originated as a partnership becomes, in effect, one and the same identity. As mentioned earlier, it is now widely thought that the cells of plants, animals and fungi arose in this way, and that their DNA-containing organelles like mitochondria and chloroplasts are derived from bacterial ancestors.

Parasitism is an extremely widespread phenomenon, which is usually viewed detrimentally, as a cause of disease and death. But is this view yet another illustration of our human tendency to draw one-sided conclusions? Could parasitism at one scale of life contribute to vitality at another scale?

Both parasitism and mutualism are concepts based on the prevalence of a 'cost-benefit' approach to classifying symbiotic relationships, which is very revealing of underlying assumptions. For example, a common schema based on this approach classifies associations between two organisms into six categories depending on whether the outcome for each organism is beneficial (+), detrimental (−) or neutral (0): so we have ++, +−, +0, −0, 00 and −possibilities.

Quite apart from the enormous difficulty of calculating what the net costs and benefits to each might actually be in any particular case, this approach confines its attention (like Newton did in his analysis of solar system dynamics) to two bodies at a time within a discrete frame of reference. It therefore avoids the complexities of accounting for the simultaneous mutual influence of three or more bodies (the 'three body problem'). Moreover, it is purely transactional in that it envisages the exchange of some kind of currency between two primarily isolated entities. It hence effectively ignores the complex dynamic neighbourhood, of which these entities are inseparable inclusions, and so may seriously misrepresent their role in ongoing interdependent natural processes. For example, we may take a limited snapshot view of a powdery mildew fungus growing on one of our crop or garden plants as a 'pathogen' 'attacking' the plant and thereby feel compelled to rally to the defence of the 'victim' by spraying it and its neighbourhood with fungicide or manipulating its genome. This may result in 'collateral damage' of the same ilk as when in human warfare, we attempt to rid a host community of its pestilential influences.

By making rash judgements based on one-sided quantitative analyses that fail to account for dynamic context, we may seriously mismanage our environmental relationships. How many of the 'diseases' that we seek to eliminate from our living space are 'diseases' of our own making? How much worse might our eliminative control measures make the situation? How many of us recognize the wisdom of Louis Pasteur's deathbed confession: 'Bernard avait raison; le microbe n'est rien, c'est le terrain qui est tout' ['Bernard was right; the microbe is nothing, it is the terrain that is all'].

5.6.1 Diversity in Life Cycle Patterns Within Natural Communities

There is a widespread recognition even in conventional evolutionary biology that not all life forms rush equally to reproduce and spread themselves in advance of others. Natural communities are populated both by 'hares' and 'tortoises'—creatures that sprint through their lives and other creatures that sustain a more durable potential in the longer run. The former kinds of creatures have been described as '*r*-

selected' and the latter as 'K-selected', in accordance with what is known as the 'logistic equation' of population growth (see Chap. 3).

It is the tortoises of this world that show us there is more to sustainable life than a relentless 'rat-race' with others. But despite the supposed influence of the Galapagos on his thinking, neither Charles Darwin nor his many adherents seem really to have appreciated this as they have persisted in defining evolutionary 'fitness' in terms of rates of reproduction. Consequently there is a perception that there is a need to reproduce and compete all the more intensively as population numbers increase and resource availability decreases. This perception is, however, incompatible with the slowing down and resource pooling that actually occurs in natural communities under such circumstances and is needed for sustainability.

In natural communities there is often a 'succession' from rapid-growing, fast-spreading, short-lived life forms in early phases of development following a local disturbance or enrichment, to more durable life forms in later stages. This can be seen, for example, in sand dunes, disturbed soil, in newly emerged volcanic islands and at the margins of lakes and pools. It leads to the formation of the grasslands, moorlands, heathlands and forests that comprise the main terrestrial communities of distinctive parts of the world—depending on the different climatic conditions prevailing at different altitudes and latitudes. These communities tend to become more complex in their structure and diverse in their composition as they develop. Innumerable life forms come to live and grow within, upon and alongside one another over diverse spatial and temporal scales in dynamic evolutionary neighbourhood.

These transformational processes occur both within and amongst the diverse life forms as they attune with their changing circumstances by varying the permeability, deformability and continuity of their dynamic boundaries. In early phases, the emphasis is on proliferation within deformable, permeable boundaries. In later phases, processes of boundary-sealing, boundary-fusion and boundary redistribution enable the effective and efficient conservation, pooling and exploration for resources that sustains the community in dynamic balance.

5.6.2   *The Vitality of Death Within Natural Community Life*

Perhaps, one of the most important lessons that an appreciative understanding of natural communities has to offer us is that far from being an 'ending', death of individual living systems is *vital* to the natural continuity and evolution of life! As I illustrate in Fig. 5.3, death serves to feed, structure, protect and transform life in a vast variety of ways.

Whether it be plant, animal, fungal or bacterial life that is consumed, feeding upon it means that it dies at some stage prior to or during the process. When we take a meal we participate in the great recycling process of the dynamic neighbourhood and global 'community of communities' that has been called 'the biosphere'.

Fig. 5.3 'Opening endings'—the dying of an elm tree opens up possibilities for a diversity of expressions of new life to emerge in its wake (oil painting on canvas by Alan Rayner 1999)

The intake of energy for this process comes largely through the reception of sunlight by the solar panels of green plants and the associated combination of carbon dioxide with the hydrogen from water to produce carbohydrates, whilst releasing oxygen. This photosynthetic receptivity of green plants makes them the sunlit world's primary producers of sources of organic carbon upon which the consumer world's animals, non-green plants, fungi and many kinds of bacteria depend. As producers, green plants are known as 'autotrophic', because they are 'self-nourishing', making and consuming their own food within their bodily boundaries. Some kinds of bacteria are also autotrophic. As consumers, other organisms are known as 'heterotrophic' because they receive food directly or indirectly from other organisms, either by absorbing or ingesting it. By 'food' I mean organic chemical substances like carbohydrates, fats and proteins that are both incorporated into the living bodily structure of organisms by 'anabolic' or 'synthetic' metabolism and used as fuel in 'catabolic' or 'destructive' metabolism.

The consumption of organic compounds as fuel involves the process known as 'respiration'. This can be thought of as a controlled explosion, analogous to that in an internal combustion engine, which releases chemical energy in a form (known as adenosine triphosphate or ATP) that supports the vibrant activity of living systems. In its fullest expression, respiration involves that other product of photosynthesis, oxygen, to support the combustion of organic fuel into carbon dioxide and water— i.e. it is the *reverse* of photosynthesis. Hence, the energy coming from the fire of a dying sun generated by the nuclear fusion of hydrogen into helium, is converted through photosynthesis into organic fuel, whose energy is released in a more earthly kind of fire, respiration, which re-produces the ingredients for photosynthesis.

This sets the scene for the global recycling scheme of living and dying known as the 'carbon cycle', in which oxygen plays a vital role due fundamentally to its spatial receptivity or 'attraction' for electrons. In many ways, oxygen can be thought of as the living world's first and deepest addiction—a substance incorporated into the very substance of organic life, which both hugely energizes and destroys living form. As oxygen receives electrons, in the course of its chemical 'reduction' to water, highly reactive intermediates are produced that have the potential, if not contained, to break down the chemical integrity of living cells as well as the fuel that these cells supply or are supplied with. The presence of these reactive intermediates causes the condition known as 'oxidative stress'.

Within the carbon cycle, complex arrays of feeding relationships make it possible for death to redistribute energy from one form of life to another. These arrays are commonly referred to using such definitive terms as 'food webs' and 'food chains', which belie their fundamental flow geometry. Within this geometry, herbivorous animals consume plants. Carnivorous animals consume the meat from other animals, both as carrion feeders and as predators. Larger carnivores consume smaller carnivores. Carnivorous plants, like Venus flytraps, supplement their photosynthetic diet by consuming small animals like flies as a source of nitrogen. Carnivorous fungi consume small animals like nematode worms. A host of small animals consume the detritus from larger animals and plants. Fungi and bacteria play enormously important roles in decomposition of the remains of other organisms as well as in parasitic and mutualistic symbioses of the kind I described earlier.

There are also many ways in which death enables redistribution from redundant to active phases of development within the same life form. A good illustration of the re-distributive role of death in the life of plants can be gained from that supposedly great symbol of resilience, a mature oak tree!

Notwithstanding its robust outward appearance when viewed from a distance, closer inspection of an oak tree reveals tell-tale signs of an ever-dying story everywhere within and upon its bark-encrusted surfaces. Imagine for a moment what this tree would look like if it had retained all the branches and leaves that it produced over its long life span: an impenetrable thicket! To attain its mature shape, the tree has undergone annual cycles of expansion and shedding of its canopy, which we can trace in the scars of detached leaves, bud scales, acorns and twigs along its branches. These detachments will have fallen to the underlying ground and been incorporated into soil through the process of decomposition, whence the mineral nutrients they contain can be transferred back into the tree through its roots and mycorrhizas. Meanwhile, any soluble sources of carbon they contain will have been transferred back into the tree via an abscission zone before fall.

It Is not only the small twigs and leaves that die and detach from the tree. Larger branches, many metres long can also succumb as the canopy expands. These often can remain attached for many years as antler-like ornaments. Their ridged and grooved sculpturing is the product of tannin-rich interfaces produced by oxygen-induced cell death in the regions between non-decayed, water-conducting sapwood and non-conducting, decaying sapwood. The tannin-rich 'heart of oak' is also produced through the death of cells in wood that has ceased to conduct water due to cavitation—the production of gas bubbles in its pipelines. Once removed from the tree, this heartwood provides a very durable timber, but within the tree it is susceptible to decay by fungi that can tolerate the tannins and carbon dioxide-rich regime to be found there. Correspondingly in many mature trees, the 'heart of oak' is actually a hollow heart, a cavity that provides a habitat for many other forms of life, and into which the tree may itself root and form mycorrhizas.

In animals, the process of metamorphosis involves the conversion from a larva to an adult, e.g. the transformation of a tadpole into a frog or a caterpillar into a butterfly. In the case of a tadpole, the tail and gills which are appropriate for a life in water degenerate and become replaced by the legs and lungs that enable frogs to make their way on land. The degeneration and re-absorption of the tail is a re-distributive process that involves what is known as 'apoptosis', developmentally 'programmed' cell death. Degenerative processes are even more apparent during insect metamorphosis, where virtually the entire muscle system of a larva is absent from adults emerging from a pupa.

Trees illustrate how death can structure life. They consist of a bark-covered set of woody channels that connect the photosynthetic canopy with its water and mineral gathering roots and mycorrhizas below ground. Both wood and bark are the products of oxygen-assisted cell death, associated with the formation of relatively impermeable compounds known, respectively, as lignin and suberin. Hence, the living (in the sense of metabolically active) tissues of a tree are distributed very thinly indeed within and over the skeletal lining that they continually add to. In somewhat similar ways, animals may fashion internal or external staging to live within and upon, both individually, as in shells and skeletons, and collectively as in coral reefs.

Another kind of staging, where programmed cell death that supports the life of a particular kind of organism serves ultimately to provide host space for a rich community of others is found in the bog-building moss, *Sphagnum*. Here, apoptosis produces a matrix of large, empty cells with porous, spirally thickened walls, interlaced with a network of narrow, photosynthetic cells. The empty cells enable the moss to be like a sponge, capable of holding up to 20 times its own body weight of water and gradually to convert initially open water into a build-up of vegetation in which other plants can take root. The dead remains of the moss form peat, which accumulates into a layer deep enough for other plants to establish.

More recently, it has been recognized that programmed cell death limits the proliferation of cells that produces cancers. Indeed cancers can be thought of as forms of life that bring death to the corporate bodies that they inhabit, and so to themselves.

Death can also deny access to the host space of an organism by potentially disruptive intruders. Both the immunity systems of animals and what are known as the 'hypersensitive' systems of plants involve the oxygen-assisted 'suicide' of host cells as a way of sealing off their bodily interiors and releasing toxic compounds that can destroy or arrest the development of colonizers. Similarly somatic incompatibility systems, of the kind that I mentioned earlier, can prevent genetic domination and loss of diversity in populations of the same species.

5.7   From Biosphere to 'Cosmosphere' and 'Quantumsphere'

Descriptions and discussions of life on Earth generally stop beyond that realm of curvature that has been called the biosphere and beneath the organizational scale at which we recognize atoms and molecules. But if we truly wish to comprehend the origins and reach of living form and pattern, it makes no sense to limit our enquiry in this way. Indeed, it is when we extend our enquiry that we can begin to recognize that the relational roots of biological behaviour can be found in what might be called 'quantum life' (cf. Marman 2016) and feed into the branches of 'cosmic life'. The findings of quantum mechanics cannot be understood objectively, but instead reflect the intangible possibilities of the relationship between space and energy that manifest in tangible material form.

In fact, it is at the scales of very large and very small that the mutually inclusive relationship between receptive space and informative energy is most evident. Simply to recognize that this relationship exists in all natural bodies does not in itself have anything to say concerning whether this relationship has always existed or whether it emerged in some way from a 'primordial soup' of undifferentiated 'energy-space'. The question I ask in this respect is 'what would Nature be like if space and energy were independent/mutually exclusive from one another?' The answer that comes back is 'utterly incoherent' (i.e. truly 'random' as per a maximal entropy state). Coherent form can only begin to emerge from this soup when local (ostensibly dimensionless) 'centres of energy' start to form cyclic relationships with one another that give rise to a 'natural, energetically enveloped body', with 'intra-space' and 'trans-space' included in (and continuous with) 'inter-'and 'extra-space' (cf. Fig. 4.4). Now, the 'life and evolution of the universe' can well and truly be under way. BUT this differentiated 'embodied form' can also 'de-differentiate', through passing its energy on elsewhere (as in 'dying'). So, perhaps, at a quantum scale we can begin to glimpse how 'life and death' could mediate a continuing relationship between coherent flow-form and incoherent energy-space, which enable the continual reconfiguration of life into myriad expressions.

In much the same way that bodily boundaries of smaller scale have been treated as the limits of discrete forms of life from cells to ecosystems, the edge of the biosphere marks where Earthly life can be mentally *alienated* from the Cosmos and made a totality in its own right. But where is this barrier that seals us off from the Heavens? One thing we can be sure of is that if such a barrier did exist, life as we know it in all its evolutionary dynamic complexity could not. Life on Mother Earth, as some like to call her, depends on the life of the Cosmos, whether or not the latter can assume organic form anywhere else in its myriad constellations. As a responsive *source* of life, Mother Earth is necessarily and simultaneously a receptive *sink* for energy flow conveyed like incoming sperm in shafts of sunlight. She cannot keep herself to herself, rotating independently about her own axis, but is inextricably caught up in the flow of inseparable cosmic inclusion of electromagnetic wind within gravitational fields that swirls, ripples, streams, connects and pools everywhere. In our human longing not to be alone, we dream of life on other planets and of breaking the bounds of Earth's biosphere to navigate our way amongst and beyond the far reaches of the solar system. But whatever was it that made us feel alone in the first place?

References

Dawkins, R. (1989). *The selfish gene* (New ed.). Oxford: Oxford University Press.
Dawkins, R. (1995). *River out of Eden: A darwinian view of life*. Basic Books.
Dowson, C. G., Springham, P., Rayner, A. D. M., & Boddy, L. (1989). Resource relationships of foraging mycelial systems of *Phanerochaete velutina* and *Hypholoma fasciculare* in soil. *New Phytologist, 111,* 501–509.
Marman, D. (2016). *Lenses of perception—A surprising new look at the origin of life, the laws of nature and our universe*. Ridgefield, Washington: Lenses of Perception Press.
Rayner, A. D. M. (1997). *Degrees of freedom—Living in dynamic boundaries*. London: Imperial College Press.

Chapter 6
The Influence of Core Beliefs and Perceptions on Human Cultural Diversity and Governance

Abstract Abstract and natural perceptions of life patterns affect human relationships and social organization in different ways. Abstract thinking is prone to isolate humanity from its natural source of life and love, through its disregard of individual receptivity. This is the root cause of the estrangement from one another and our natural neighbourhood that we experience in our modern world, leading to needless ideological conflicts and environmental devastation, as well as profound scientific and philosophical misunderstanding. Abstract logic and language reinforces beliefs in independent self- and group-identities, external authority, competitive success, strength as good, weakness as bad, money as wealth, time as an external occurrence, unity as an antidote to division and death as finality. These beliefs are transformed into more realistic, loving and sustainable understandings through awareness of natural inclusion—but only if people are psychologically receptive enough to relinquish the definitive assumptions that have become deeply entrenched into human thought for millennia.

6.1 Human Estrangement from Nature

In this chapter, I will explore how the way we perceive the origin of life patterns affects human psychological and social development. I will outline why abstract perceptions of space, time and boundaries give rise to many deep, interrelated problems of living that could be remedied by an appreciation of natural inclusion. To do full justice to these problems and their origins would require much more in-depth treatment than is possible here. My aim therefore is simply to bring them to readers' attention for further consideration.

I will begin by saying how extraordinary it has always seemed to me that the human psyche should ever have believed itself to be separate from Nature. And yet that belief is exactly what abstract perception sustains, reinforced by definitive language and thinking. The distancing of observer from observation has the effect of psychologically isolating humanity from its source.

© The Author(s) 2017
A. Rayner, *The Origin of Life Patterns*, SpringerBriefs in Psychology and Cultural Developmental Science, DOI 10.1007/978-3-319-54606-3_6

The fact that nothing is independent from its environmental context is what Nature—from quantum to galactic scales—is continually trying to tell abstract science: 'Not only are you using the wrong lens to study me, but as a consequence you are also using the wrong logic and the wrong language to try to understand and describe me'. It does not make natural sense to view our world solely through an objective lens, to use logic that excludes space from matter and to use language that imposes definitive boundaries on fluid forms (cf. Marman 2016).

The mother and father of all natural patterns, processes and relationships is the mutually inclusive relationship between space and energy. Space is a receptive presence everywhere and energy is a locally informative presence. From this relationship, the vast variety of material form emerges. To deny this primal relationship renders rationalistic thought into a purely materialistic conception without ancestry. The orphaned thought stumbles around its misconceived material world in a way that engenders profound paradox, misunderstanding, conflict and suffering. It comes up with all sorts of estranged beliefs concerning what it perceives as its independent, free-willed existence in which it makes conscious choices of its own that have nothing to do with the dynamic origins of natural energy flow. This orphaned thought enshrines and perpetuates these beliefs in its philosophy, mathematics, science, theology, politics and education systems. Then it looks for a scapegoat to blame for its problems, instead of diagnosing the root of these problems within its abstracted self.

Such is our modern human predicament: that we have allowed beliefs rooted in abstract perception and thinking to override our natural sense and experience. We have lived these beliefs to the point that our lives contradict our true human nature as needful, life-loving creatures. We are inescapably dependent on our habitat and one another for the food, fuel, water, medicine, shelter, cleansing, clothing, equable temperature and love that sustain our lives. Yet we deny this reality. Despite our burgeoning technological prowess, we live ill-at-ease with one another and our surroundings. Some of us are aware that something is fundamentally wrong with our thinking, but very few are aware of what this is and how to remedy it by dispelling the illusion of our separation from Nature (Hutchins 2014). We worship at the altar of unrealistic ideals, beliefs and values and teach our offspring to do the same, while ignoring what truly sustains life and love.

6.2   Belief in Independent Identities

I recall my parents once being congratulated for providing the genes that accounted for my early academic success!

The abstract perception that our self-identity is genetically determined at conception, and remains fixed thereafter regardless of what circumstances we find ourselves in, sets us up to be categorized and judged as performing objects. We then feel—and are deemed by others—to be solely responsible for the way we behave. We take credit for our successes and are blamed for our failures. Pride, shame, guilt and anxiety are prominent emotions. These emotions profoundly shape our behaviour and relationships with one another. We have become driven primarily by fear and lust, not love and understanding.

Seeing ourselves as openings, centres of receptive stillness around which energy circulates, transforms our perception of self- and group-identities. It shows us that who we are actually arises from the inseparable, mutually inclusive relationship between us and our environment. This aligns with our actual experience of life, living within the influence of our natural neighbourhood, does it not? Our identities continually transform as we experience and learn from life, much as the character of a river changes as it shapes and is shaped by a variable landscape. Such developmental plasticity is in fact crucial to the ability to sustain life in a heterogeneous and changeable environment. All life forms change their pattern of expression in differing circumstances, as is beautifully illustrated by fungi (see for example, Chap. 3, Fig. 3.3; Chap. 4, Fig. 4.3).

This receptive–responsive awareness can be suppressed, however, if our belief in independent agency is strong enough. And there is a very powerful way in which this belief is enforced upon us by an intimidating form of governance of 'power-over' instead of 'power-with'. This intimidation imposes an unnatural order upon our lives through the belief that in the absence of prescriptive 'rules of conduct', our social coherence would break down into an anarchic free-for-all instead of mutual complementation.

6.3 Belief in External Authority

How many of us, whether we freely admit it or not, live our lives in fear of some kind of external Executive Body that independently controls and judges us, rewarding what it approves and punishing what it rejects? This profound fear is intensified rather than released by the prospect of Death, if this prospect comes with a threat-and-promise of an eternal afterlife of torment or serenity depending on how well we behave during our limited lifespan on Earth. This threat-and-promise does not, however, guarantee 'good behaviour'. Rather it is a source of the most profound cruelty that we humans are capable of inflicting on ourselves and our natural neighbourhood. This cruelty is being acted out blatantly and globally as I write. It places us in a 'Catch 22', torn between whether to serve our individual or group interests, and instructed that it is 'good' for us both to compete and cooperate with others depending on whether they are deemed 'good' or 'bad'. We speak of the desirability of 'Peace', while arguing the case for a 'Just War' and denigrating conscientious objection to this case as 'cowardice', or worse. The great irony in all this is that the very idea of external Authority is perpetuated. It creates a false dichotomy between material and immaterial presences due to a misperception that only matter is real. Space is deemed not real and discounted. In reality, space and matter are distinct but mutually inclusive.

Belief in external authority is closely associated with the abstract idea of a Great Chain of Being or Hierarchy, derived from the philosophy of Plato, Aristotle, Plotinus and Proclus. This strict religious 'Order of Merit' reaches down from the utter superiority of an *externalized* God and cast of Angels to the dominion of Superior Man over Inferior Man over Inferior Nature. It still holds a powerful sway over our human attitudes towards one another and our natural neighbourhood. It even persists in Darwinian 'selection' as the preservation of favoured races in the struggle for life: an external agency views Nature as a set of discrete objects and decides which ones are good or not good enough to survive.

Symbolically, the story of the New Testament comes across to me, however, as that of an external Authority Figure come down to Earth to be a Lover amidst the flow of life instead of an independent and manipulative arbiter. In this new-found situation, a radical shift in attitude takes place, from one of judgemental condescension by a singular figure to one of tender care for the needs of self and others. This care is held in the natural companionship of their natural neighbourhood. A top-down hierarchy of decreasing perfection is subsumed by the compassionate recognition of mutual needfulness and vulnerability. Self, according to the New Testament lens, naturally becomes receptive and attentive to what is needed to sustain life, by way of attuning with the availability and flow of sources of energy. Abstract Superiority is replaced by the Natural Communion of Each in Other's embrace.

As I described in Chap. 4, Darwinian selection simply replaces the overarching power of God as an external ordering influence, with Nature as a judge of itself. This gives rise to 'Just So' stories, which tautologically explain away all the attributes of living creatures as the product of competitive adaptation to a pre-existing set of environmental specifications, a so-called 'niche' or 'box of limited space'. Tales emerge of selfish genes as randomly generated units of selection and monkeys randomly generating the works of Shakespeare on a typewriter given infinite time, all couched in binary code of 1 or 0 (cf. Dawkins 1995).

So how, then, can an appreciative understanding and awareness of natural communion help to transform the way we have become prone to think about the evolution of natural diversity and our human place in it? How can we live co-creatively, sustainably and compassionately without feeling forced to compete or cooperate with or against one another? How can we recognize the enormous variety of life-sustaining relationships that can develop between different neighbouring identities besides unity or division?

Quite simply, we can STOP thinking of ourselves and others as autonomous, free-willed objects subjected to external administration and judgement! And we can START thinking of ourselves and others as dynamic inhabitants and expressions of our natural neighbourhood, living within each other's mutual influence. We may then understand evolution as a cumulative, fluid process of mutual transformation within each other's company, not a selective sorting mechanism that judges goodness of fit to a preconceived ideal.

In other words, we need to move on from viewing evolution in terms of abstract selection by an extrinsic arbiter, to understanding evolution as an intrinsic process of *natural inclusion*, the fluid-dynamic, co-creative transformation of all through all in receptive spatial neighbourhood. Once we make that move and embrace it in our minds, hearts and guts, the restrictive way of life that has been blighting our humanity for millennia radically transforms into one that makes a lot more natural sense. Instead of denigrating our natural companions as inferior, we recognize their vital place in the natural scheme of life, where it truly does take all kinds to co-create a viable, sustainable world.

> Natural communities, like forests, don't require intervention by a supervisor to ensure their lively, sustainable functioning as coherent evolutionary systems – why should we?

6.4   Belief in Competitive Success

The idea, deeply embedded in Darwinism, that life is a competition between opposing entities in which success and failure are defined by 'winning' and 'losing' is an invention of abstract perception. As discussed in Chap. 4, it impedes rather than drives evolutionary innovation and diversification. It precludes any possibility of co-creativity and so is ultimately self-defeating. To suppose otherwise is a huge mistake. In human communities, win-or-lose competition is a source of enormous waste, conflict and misery, not flourishing. Competitive 'triumph' has nothing to do with the participatory reality of evolution in natural living communities. Competitive success indicates excellence only within pre-defined limits and offers no guarantee of continuation when circumstances change. Evolution is a learning process of cumulative transformation, not running ever faster on the same spot like a demented Red Queen. History teaches this lesson repeatedly: what serves well in one context ceases to do so in another—light bulbs supersede candles as we learn how to channel electricity through copper wiring.

So, if win-or-lose competition impedes evolutionary possibility, what naturally opens it up? The answer is simple: the openness to possibility that is implicit in natural fluid geometry!

Here I want to distinguish carefully between abstract win-or-lose *competition* and the *rivalry* that is evident in natural territoriality and incompatibility. There is a radical difference in intention between that which desires primarily to 'win', and that which seeks primarily to serve its needs as well as it possibly can in circumstances where it finds itself accompanied by others doing likewise. Rutting stags are not 'in it to win it', they are drawn into rivalry with others by the receptive influence of the does. Adjacent fungal colonies are no more intentionally competing against one another to see who is best, than adjacent river basins on either side of a watershed.

Abstract competition is ruthless in its treatment of what are considered to be its opponents, and is ready to use any means it can to serve its purpose: 'all's fair in love and war' is its maxim. There is no place in its armoury for contemplation of its actual situation in relation to its neighbourhood, beyond its determination to predominate. Rivalry, on the other hand, is deeply receptive and responsive to the needs and distinctive qualities of itself and others within its neighbourhood, to which it can contribute and from which it can learn as a *participant* in life. Rivals can and sometimes do become partners, as in the formation of mutualistic symbioses described in Chap. 5. In a natural world where life and evolution is sustained by natural energy flow, the capacity to yield to others is just as vital in the overall scheme of things as the capacity to gain from others.

6.5   Belief that Strength Is Good and Weakness Is Bad

Strength is widely regarded as essential to the success of human organizations and individuals. By the same token, weakness is commonly held responsible for organizational and individual failure and breakdown. When things start to go wrong, the call therefore almost invariably goes up for the strength needed, often in the guise of some kind of superior person or 'strong leader', to put things right.

The kinds of qualities expected of a strong individual are, however, liable to depend greatly upon whether the underlying perception of success in life is abstract or natural. Abstract perceptions are based on the imposition of order and predictability onto what would otherwise be expected to be shambolic. Authoritarianism—Iron Rule imposed by Iron Men and Iron Ladies is therefore called for, and all-too-often provided, with what can prove to be catastrophic and long-lasting consequences that bring immense human suffering in its wake. The rise to power of all sorts of despotic figures, throughout recorded human history is testimony to this tendency.

The simple truth is that no individual can ever be all-knowing and free from weakness, so placing one's faith in someone who claims or is expected to be supreme is unrealistic. The essence of true democracy is the bringing together of diverse abilities and viewpoints that combine into comprehensive awareness and action: governance by all, through all for all. The selection by election of governing elites supported by the majority of the people is not true democracy. It is the suppression of the minority administered by the most popular, a sure route to corruption and dishonesty in the quest for power, which is played out continually in modern politics. Majorities do not have a monopoly on truth, especially when they can be persuaded to think partially instead of comprehensively. The pretence of democratic 'strong government' can be just as corrupt and dishonest as the dictatorship that it claims to replace, if not more so.

This is not to suggest that democracies should not have leaders, but more to recognize that our idea of strong leadership needs to change from that based on authoritarian rule. Natural leaders are not divisive rulers. They are local pioneers, guides and coordinators who enable others to explore, learn and live together co-creatively.

Ecosystems such as forests exemplify natural democracy as coherent, evolutionary organizations without any overarching sovereign authority—the 'Lion King of the Jungle' is a fiction—but are not without the intrinsic, orchestrating influence of those amongst their membership. A tree does not impose its authority on its neighbourhood, but its receptive presence brings a host of organisms into life within its vicinity. Similarly an inspiring teacher does not lay down the law, but provides experience, knowledge and example as guidelines for others to follow. And 'pioneer' life forms open the way for others to follow in natural succession.

Meanwhile, not only is individual weakness inevitable, it is actually a vital quality of evolutionary flow form, which facilitates exploration of diverse possibility in natural companionship. The art of natural creativity and diversity arises through the mutually inclusive relationship between informative flux and receptive space, as I sought to illustrate in the painting shown in Fig. 6.1 and associated poem, 'Holding Openness'.

Fig. 6.1 'Holding Openness'—light as a dynamic natural inclusion of darkness continually brings an endless diversity of flow form to life (Oil painting on canvas by Alan Rayner 2005)

Holding Openness

You ask me who you are; To tell a story you can live your life by; A tail that has some point; That you can see; So that you no longer; Have to feel so pointless; Because what you see is what you get; If you don't get the meaning of my silence; Because you ain't seen nothing yet

You ask me for illumination; To cast upon your sauce of doubt; Regarding what your life is all about; To find a reason for existence; That separates the wrong; From righteous answer; In order to cast absence out; To some blue yonder; Where what you see is what you get; But you don't get the meaning of my darkness; Because you ain't seen nothing yet

You look around the desolation; Of a world your mined strips bare; You ask of me in desperation; How on Earth am I to care?; I whisper to stop telling stories; In abstract words and symbols; About a solid block of land out there; In which you make yourself a declaration; Of independence from thin air; Where what you see is what you get; When you don't get the meaning of my present absence; Because you ain't seen nothing yet

You ask of me with painful yearning; To resolve your conflicts born of dislocation; From the context of an other world out where; Your soul can wonder freely; In the presence of no heir; Where what you see is what you get; When you don't get the meaning of my absent presence; Because you ain't seen nothing yet

You ask me deeply and sincerely; Where on Earth can you find healing; Of the yawning gap between emotion; And the logic setting time apart from motion; In a space caught in a trap; Where what you see is what you get;

And in a thrice your mind is reeling; Aware at last of your reflection; In a place that finds connection; Where your inside becomes your outside; Through a lacy curtain lining; Of fire, light upon the water

Now your longing for solution; Resides within and beyond your grasp; As the solvent for your solute; Dissolves the illusion of your past; And present future

Now your heart begins to thunder; Bursting hopeful with affection; Of living light for loving darkness; Because you ain't felt no thing yet.

6.6   Belief that Money Is Wealth

'Money can't buy me love', the Beatles once sang—rather ironically in view of the fact that singing about love brought them loads of money! Since that time, in the 1960s, when 'Love' was top of the agenda for discussion, and materialism was seriously being questioned, albeit with the hallucinogenic aid of 'Lucy in the Sky with Diamonds', modern cultures have reverted to an obsession with money and power as primary motivating forces. Under the influence of 'Game Theory', neo-Darwinian 'sociobiology' (incorporating 'selfish gene theory') and mone-tarism, the socio-economics of restrictive self-interest and purely financial wealth prescribed by Adam Smith have returned to prominence, and love has disappeared from the map of serious and even polite discussion. Disparities between rich and poor have grown even wider, and the need to sustain economic growth continues to take precedence over environmental sustainability, for all the talk about the need to protect and enhance our quality of life. Meanwhile public knowledge and appreciation of biological diversity within their own neighbourhoods continues to dwindle as our educational institutions become dominated by commercial interests instead of seeking to enhance public awareness of the natural world as it actually is.

The fallacy in prioritizing financial wealth resides in the fact that money as an abstract commodity based on the principle of personal ownership of property and promise of return in the future is NOT natural currency: energy is natural currency.

There is clear anthropological evidence that prior to barter and financial transaction, human social organization was and in some indigenous communities still is primarily orchestrated according to principles of 'gift flow'. These correspond closely with the circulatory and redistributive supply, receipt and temporary retention of natural energy flow characteristic of natural ecosystems. Life is appreciated as a gift of natural energy flow, to be received, cared for and passed on in the goodness of place-time. Even in modern cultures, intangible qualities of love and artistic creativity are a shared source of profound human pleasure and caring that defy, and are defiled by, any attempt to commoditize or quantify them.

At the heart of traditional gift flow is *trust* in the principle that what is freely given is equally freely returned in the long-run, such that whoever gives away most also receives most, and vice versa. This harmonizing principle is broken as soon as anyone accepts without giving or vice versa. Such restrictive practices give rise to a breakdown of trust, which in turn gives rise to further restrictive practices, setting the scene for a vicious cycle of competition, conflict and increasingly rigorous legislation to define trading practice and monetary transactions. By the same token, such restrictive economic rationality is associated with the localization (privatization/nationalization) of self and/or group identity and individual or public 'rights' of property *ownership*. Sometimes systems of gift flow may operate within family/social groupings alongside rigidly structured trading or economic practice between groups. This implies a hard boundary limit between the two and a resulting 'double standard' sometimes referred to as 'the double law of Moses', which permits repayment of a loan to be demanded from 'another', but not from a 'brother'. In other words, there is one rule for 'insiders' or 'familiars' and another for 'outsiders' or 'strangers' (Hyde 2006). The question then arises as to where and when to define the limit between one and the other. Where and when does the natural inclusion of each in the other's gift end, and the abstract estrangement and exploitation of the other's needs and talents begin?

Having recognized the origins of financial systems in abstract notions of personal property, future promise and fear and exploitation of strangers, it is easy to see how these systems become a source of human conflict and disparities in social status that have little to do with individual and collective contributions to natural energy flow and quality of life.

6.7   Belief in Time as an External Occurrence

Economic competition in modern human societies and organizations is closely related to another source of enormous stress and wastage of energy: *busyness within a restrictive time frame*. This arises from the abstract notion that superiority of performance increases as the time taken to complete an action decreases. A common expression of this is 'time is money', whereby time, like money is perceived as an abstract commodity,

> The fallacy here resides in the fact that time as an external occurrence independent of space and energy does not exist.

Our natural perception of time arises from our implicit, if not explicit awareness of living. Like the spinning, sun-orbiting planet we inhabit as dynamic rotational energetic inclusions of place-time—we experience the continuous *pulse and circulation* of natural energy flow, in natural rhythmicity, including the alternation between daytime and night-time.

The abstract perception of time as a measurable commodity, divisible into discrete units or intervals (seconds, minutes, years etc.) arises as a derivation from the flattening out of this circulation into a straight line. The resultant 'arrow of time', as a 'fourth dimension', stretches from past into future via an eternally shifting definitive cut-off point between the two, which is called 'the present'. The use of this arrow of time as a baseline against which to measure performance always begins with this non-existent cut-off point in the present and ends with another non-existent cut-off point 'sometime and somewhere in the future that has yet to arrive'.

As mentioned in Chap. 3, Newton's 'Laws of Motion' and associated invention of calculus are based on this flattening out or 'linearization' of continuous dynamic curvature into local flat lines. These local flat lines, known as 'infinitesimals', provide a useful way of calculating (but in reality only simulating by approximation) the trajectories of what are treated as if they are independent objects forced into motion by externalized agency. As a calculating tool, this is unproblematic. The problems begin when we start treating ourselves and others as if we really are such independent objects.

Abstract perceptions of time rule out the rotational liveliness that occurs over spatial scales from subatomic to galactic, into the local deadlines and external force that we have allowed to overrule our natural lives. Instead of simply using 'clock-time' as a convenient source of guidance, helping us to plan meetings, for example, we find ourselves trying to pack as much as we can into a restrictive schedule between enforced deadlines. We believe that this makes the most productive and efficient use of our time, especially our time in what we think of as 'the workplace'. In truth, however, it is soul-destroying, wasteful and counter-creative, through its denial of our needs as living, loving inhabitants of 'place-time', our natural neighbourhood.

6.8   Belief in Unity as an Antidote to Division

The call for 'unity' to remedy the division of a human community into antagonistic factions is a corollary of the belief that an ordering influence is necessary to prevent the breakdown of social structure into an anarchic free-for-all of disparate, independent self-identities. As such, it is also closely related to the philosophy of 'holism' put forward by General Smuts (1926), based on the notion that 'the whole is more than the sum of its parts'. Instead of appreciating the vitality of natural nonconformity to the evolutionary co-creativity of natural community life, all are required to combine together as 'One', with a common purpose that overrides individual differences and aspirations. Moreover, this call most often goes out politically as a response to what is perceived to be a common enemy, for example in wartime. So small-scale hostility between individuals is replaced by grand scale hostility between groups.

The fallacy of seeking 'unity' as a basis for social coherence arises from the fact that the imposition of conformity upon individual variation produces a sterilized monoculture, not a fertile community. This fallacy can quickly be recognized through appreciating the functional dependence of natural collective organizations upon the diverse contributions of their membership. For example, unlike plantations, natural forests are not fenced-in monotonous stands of the same kind of tree. They are rich assemblages of all kinds of trees, herbs, fungi, bacteria and animals of different size and age, each of which influences the lives of the others in diverse ways to produce a dynamic collective that lives openly attuned within and as a co-creator of its environmental context. Human communities originally evolved and functioned in much the same way. All that prevents them from continuing to do so is the abstract division of different kinds into definitive categories, whereupon the scene is set for trenchant opposition between them in place of co-creative interplay. This opposition cannot be remedied using the same kind of 'one or many' thinking that gave rise to it. It can only be remedied by resisting the compulsion to divide or unify and recognize instead that:

> United we stall, divided we fall; dynamically diverse we thrive in the energy flow of each in all.

6.9   Belief in Death as Finality

As I discussed in Chap. 5, death is actually vital to the evolutionary sustenance of life as an expression of natural energy flow. Death feeds, structures, protects and transforms life as a gift of natural energy flow to be received, sustained and passed on in natural relay, not a mad dash to be first past the finishing post.

Take another look at the interdependent relationship between explorative 'pulse' and integrative 'circulation' evident in the example of 'fungal foraging' shown in Fig. 4.3. Notice how renewed exploration is made possible through redistribution from dying initial growth, via a persistent network of anastomosed mycelium. A fully integrated—and hence retentive system is incapable of such renewal. It lacks 'pioneers' and so becomes 'self-limiting'—'all past memory and no future imagination'.

Abstract human perceptions of discrete boundaries marking the beginnings and endings of life have, however, reinforced the belief that death is an ultimate discontinuity, which finally cuts life off from its past or future. Faced with this apparent finality, life appears as a fixed term contract that begins and ends in either nothing or eternity, which humans have to make the best of in one way or another. How does this affect the way we live our lives? Is it best to forget about past or future in order to focus all our efforts on enjoying what is current? How, then can we care for and learn from our ancestry or prepare a sustainable path for our offspring? Do we resist change by clinging to the knowable past? Do we sacrifice our present life to what we perceive as a desirable future, assuming this can accurately be forecast?

Perhaps one of the most devastating consequences of regarding death as finality is the notion that it represents some kind of retribution for failure to live as one could or should. This can easily lead to moral judgments upon others' rights to live that fuel ideological conflicts and to the great injustice of abstract historical reviews and Darwinian retrospection that depict dying out of lineages as the result of failure to succeed in the struggle for life. The truth is that all life's endeavours contribute energy to the flow of the mainstream and deserve to be acknowledged.

6.10   From Abstract Rationality to Natural Inclusion: Evolution, not Revolution

Some readers of this book, and especially this chapter, may perceive it as a call for revolutionary overthrow of all that abstract thought and objectivistic science has led us to believe about ourselves and the natural world, and its replacement by the new worldview of natural inclusionality. I do not see it that way, for the same reasons that I do not see extinction and replacement as the basis for evolutionary change. What I am calling for is an evolution of thought based on a truly impartial evaluation of what the imaginary dissociation of material from immaterial presence has led us to believe in, and how this could be *transformed* into more realistic understandings through awareness of natural inclusion.

At the core of this evolution is a call for dynamic *balance*, not polarized *one-sidedness*. The 'pendulum diagram' sketched out in Fig. 6.2 shows the difference between abstract and natural perceptions of 'self' in relation to 'environment', and how this relates in turn to different perceptions of space and boundaries. So much of our modern difficulty arises from excluding 'the middle way' that is able to resolve the incompatibility between alternative abstract views.

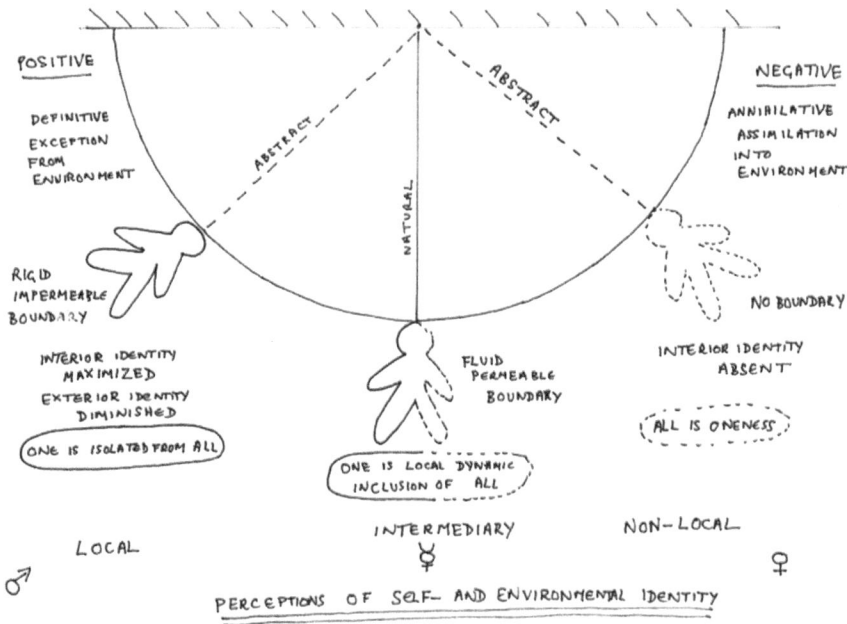

Fig. 6.2 Abstract and natural perceptions of self and environment (Freehand sketch by Alan Rayner 2015)

Instead of viewing this diagram as an outsider, you may find it helpful to imagine yourself within each of the three figures. Does the following description hold true?

Within the definitively bounded figure (LOCAL), you feel entirely alone within yourself, aware of your set location as a centre of 'your' universe but utterly detached from it: you judge everything around you against a fixed reference frame set by yourself. This is the Cartesian view of the objectivistic framework, which still holds sway in abstract scientific methodology and theory, and treats boundaries as definitive dead zones ('hard lines'), not live interfacings. This is 'separate knowing'. Within the figure with no boundary (NON-LOCAL), you feel indistinguishable from all around you, yet have no sense of locality and so 'are lost in space'. You are everywhere and nowhere at once. This is 'connective knowing', which treats boundaries as non-existent illusions. It annihilates or diffuses self-identity to all around. To some it feels like bliss, a realm of utter calm, utter passivity, lack of agency, which simply accepts whatever comes, unconditionally. Within the figure with dynamic boundaries (INTERMEDIARY), you have a sense of being included in all around you, without entirely losing your locality or agency, aware both of the omnipresent calmness of space and the local excitability of energy within your living boundaries as dynamic distinctions between your spatially continuous inner and outer worlds. This is what it truly means to be alive, involved and fully aware of your self-identity as a dynamic natural inclusion of neighbourhood.

You might also find it helpful to view the image as a photographic 'negative' of white linings on dark background. This would bring out the nature of all natural figures as natural inclusions of space (darkness) in flux (light).

Abstract thinking removes the middle ground of self-identity as a dynamic inclusion of neighbourhood. This explains why two incompatible kinds of abstract logic, have been at odds with one another for millennia. 'Two-value logic' (also known as the Law of the Excluded Middle) straightforwardly regards one or other of the two mutually exclusive alternatives (bounded or unbounded) to be 'true' and the other as 'false'. Dialectic logic holds both alternatives to be equally true, which results in paradox. Since there is no way to resolve this paradox naturally, by allowing boundaries to fluidize and space to be continuous, the brutality of one and the softness of the other are held in 'living contradiction'.

It is as though, in the attached diagram, the hard-bounded and the unbounded are pummelled like a boxer's punchball between alternative forms of abstraction, instead of being allowed to balance dynamically.

Fear of the alternative then drives violent opposition between irresolvable desires. The antipathy between each and other drives the opposition of both to what brings each into the other's embrace. The rigidly encapsulated ego sees the need for softness but cannot relax its self-definition for fear of losing its identity and mastery over other. The dissolved ego feels the need for agency but fears the potential brutality of self-encapsulation.

And so the living contradiction of each against other continues, without the natural resolution that is always possible through the mutual inclusion of receptivity and responsiveness that we find in each other's natural embrace. Abstract logic denies this natural resolution through false dichotomy. But we *can* leave this false dichotomy behind us and evolve into a more open-ended, spontaneous experience with natural wisdom. Yes, *we* can!

References

Dawkins, R. (1995). *River out of Eden: A Darwinian view of life*. New York: Basic Books.

Hutchins, G. (2014). *The illusion of separation—Exploring the cause of our current crises*. UK: Floris Books.

Hyde, L. (2006). *The gift—How the creative spirit transforms the world*. Edinburgh: Canongate Books.

Marman, D. (2016). *Lenses of perception–A surprising new look at the origin of life, the laws of nature and our universe*. Ridgefield, Washington: Lenses of Perception Press.

Smuts, J. C. (1926). *Holism and evolution*. New York: Macmillan Press.

Chapter 7
Epilogue—The Co-creative Partnership Between Responsive Flux and Infinite Grace in the Life of Every Body

Abstract The receptive presence of intangible space is central to self-identity and love and understanding of one another. It is in the heart of our 'Being' as sentient life forms, where it combines with energetic flux to co-create tangible natural bodies as continuous 'Becomings'. If it were to be personified, she could be known as Grace, that dwelling place for inner peace amidst the turbulence of everyday life.

7.1 Flux and Grace

In Chap. 1 of this book I described how my quest to understand the origin of life patterns has taken me very deep into the fundamental questions that contemplative human beings have always pondered. I invited you to accompany me, through the pages of the book, on a personal 'nature walk' with me. I wanted us first to recognize the patterns themselves, then to question what kind of enquiry is needed to understand these patterns as variations on an underlying theme and finally to consider why this understanding is so important for human beings. I wondered whether we would find ourselves, as writer and readers, arriving in the same place.

As I was writing, I was, as ever, engaged in conversations that challenged me to consider what my own core beliefs are, and how these relate to others' perceptions and spirituality. In one conversation I was asked whether I thought of the receptive stillness of space as a 'Being', and if so, how I might personify this Being. Here is my response, which summarizes where I find myself currently.

I think of space as 'infinite being'—I might even call this Being 'Infinite Grace'—which Every Body includes within its 'core' or 'heart'.

I think of Every Body—from sub-atomic 'particle' to 'Universe'—as 'Energised Being', a receptive core enveloped in the flux of 'Becoming', both permeated and surrounded by Infinite Grace. I imagine 'the Universe' as a 'Body' within the Infinite Grace of the 'Cosmos'.

I envisage Every Body as living in dynamic, mutually transforming relationship with every other Body within the Body of the Universe within the Grace of Space. This relationship results in the giving and receiving of gifts of energy and even

A. Rayner, *The Origin of Life Patterns*, SpringerBriefs in Psychology and Cultural Developmental Science, DOI 10.1007/978-3-319-54606-3_7

bodily coalescence, it does not make everything instantaneously 'interconnected'. There can be repulsion as well as attraction, but both repulsion and attraction are vital to the fluid responsiveness of the body.

'Infinite Grace' alone has no energy—She is pure receptivity, pure, intangible Stillness, but this inviting receptivity is not manifest without embodiment. There has also to be flux for bodies to form. If there were just One Body in Nature, there would still be energy, because there can be no body without energy. Energy is intrinsic within Body. In the world as we know it as human beings-and-becomings, this flux comes via relationships with other bodies, and both via light and material assimilation. Relationships provide both the potential and actual channels for flux, but this does not make them ultimate 'creators' or 'causes' of flux. Hence, I regard flux as a primary presence, along with 'Infinite Grace', both of which combine co-creatively into embodied beings expressing diverse patterns of life.

Natural Space is Infinite Grace—the Heart of True Self and Everywhere Else.

Could this be where you too find yourself arriving?

What *appears* objectively from outside to be a definitive boundary limit is *experienced* from within as a portal, an intangible zero-point of space and time, like

Fig. 7.1 'Loving error'—the heart of natural energy circulation brings positive pulsing and negative draining together through the receptivity of space here, there and everywhere (oil painting on board by Alan Rayner 1998)

that which occurs between lips opening and closing in a kiss (X) between dynamic bodily envelopes meeting and parting in a transference of energy from each into other. No life form can emerge from or fuse with another without passing through this transitional point. Real life depends on forming relationships with neighbours that enable energy to be released (as in dying) and returned (as in birthing) in continuous relay through a receptive and responsive heart: as depicted in Fig. 7.1.

Index

Note: Page numbers with *f* indicate figures

A

Abstract perception, 87, 88, 91, 92, 96
Abstract theories of origin and transformation
    of life, 47–49
Apoptosis, 84
Arid confrontation, 3, 4*f*
Atomism, 30–31
Atomistic thought, 31
Authoritarianism, 92

B

Becoming, 101, 102
Being, 101
Belief, 87
    from abstract rationality to natural
      inclusion, 98–100
    in competitive success, 91–92
    in death as finality, 97–98
    in external authority, 89–91
    human estrangement from nature, 87–88
    in independent identities, 88–89
    money is wealth, 94–95
    strength is good and weakness is bad,
      92–94
    in time as external occurrence, 96
    in unity as antidote to division, 97
Bill Yidumduma Harney (BYH), 57
Biological life, 62, 75
    as embodied water flow, 69–71
Biological survival/preservation, 50
Boolean logic, 31
Boundary-differentiating and
    boundary-integrating processes, 52–53, 52*f*
British educational system, 2
Building blocks of life, 23

C

Carbon-based biological life, 70
Carbon cycle, 83

Carnivorous animals and plants, 83
Cell growth, 73
Centaurea calcitrapa (star thistle plant), 23,
    23*f*
Certitude, sense of, 30
Circles in natural world, 38
Co-creativity, 52, 61, 101–103
Competitive success, belief in, 91–92
Completeness, 33
Complex number, 35–36
Complex plane, 36
Complexity theory, 32, 80
Conflict, 2, 75
Continuity, 31–32, 56, 59, 61
Contradiction, 1–2
Coprinopsis picacea (magpie fungus), 41, 42*f*
Cosmosphere, 85–86
Crazy paving, 42–43
Cretan liar paradox, 33
Crystalline structures, flat surfaces of, 39
Culture shock, 2, 3

D

Darkness of space, 18–19
Darwinism, 46, 91
Death as finality, belief in, 97–98
Death within natural community life, 81–84
Dendritic branching patterns, 41
Determinism, 48
Deterministic unpredictability, 33
Discrete logistic difference equation, 34
Discreteness, 45–46
Diversity, 49
    within cellular life, 71–72
    within community life (*see* Diversity,
      within community life), 78–80
    within multicellular life, 72–74
    within social life, 74–78
Diversity, within community life, 78

© The Author(s) 2017
A. Rayner, *The Origin of Life Patterns*, SpringerBriefs in Psychology
and Cultural Developmental Science, DOI 10.1007/978-3-319-54606-3